Generalized Topology Optimization for Structural Design

Yi Min Xie

Generalized Topology Optimization for Structural Design

 Springer

Yi Min Xie
RMIT University
Melbourne, Australia

ISBN 978-981-96-4523-7   ISBN 978-981-96-4524-4 (eBook)
https://doi.org/10.1007/978-981-96-4524-4

This work was supported by Australian Research Council (FL190100014).

© The Editor(s) (if applicable) and The Author(s) 2025. This book is an open access publication.

Open Access This book is licensed under the terms of the Creative Commons Attribution 4.0 International License (http://creativecommons.org/licenses/by/4.0/), which permits use, sharing, adaptation, distribution and reproduction in any medium or format, as long as you give appropriate credit to the original author(s) and the source, provide a link to the Creative Commons license and indicate if changes were made.
The images or other third party material in this book are included in the book's Creative Commons license, unless indicated otherwise in a credit line to the material. If material is not included in the book's Creative Commons license and your intended use is not permitted by statutory regulation or exceeds the permitted use, you will need to obtain permission directly from the copyright holder.
The use of general descriptive names, registered names, trademarks, service marks, etc. in this publication does not imply, even in the absence of a specific statement, that such names are exempt from the relevant protective laws and regulations and therefore free for general use.
The publisher, the authors and the editors are safe to assume that the advice and information in this book are believed to be true and accurate at the date of publication. Neither the publisher nor the authors or the editors give a warranty, expressed or implied, with respect to the material contained herein or for any errors or omissions that may have been made. The publisher remains neutral with regard to jurisdictional claims in published maps and institutional affiliations.

This Springer imprint is published by the registered company Springer Nature Singapore Pte Ltd.
The registered company address is: 152 Beach Road, #21-01/04 Gateway East, Singapore 189721, Singapore

If disposing of this product, please recycle the paper.

Dedicated to Shuiqing Zhang

Preface

There are many assumptions commonly used in topology optimization for structural design. These assumptions include pursuing a unique, globally optimal solution—the 'best' design; conducting optimization within a predefined design domain, under prescribed load conditions and predetermined support conditions; and focusing on structural performance without considering aesthetic preferences of the designer or client.

This book challenges conventional assumptions in structural topology optimization. Through a systematic discussion and numerous examples, we demonstrate that these assumptions are not only unnecessary but also impose severe limitations on design freedom and hinder creativity in structural design.

The generalized topology optimization framework advocated in this book empowers topology optimization techniques to find more practical and desirable solutions to a wide range of structural design problems.

The findings and methods presented in this book are the results of collaborative research efforts undertaken with my remarkable colleagues, including many Ph.D. students and postdoctoral researchers, primarily since 2019. I wish to thank everyone who has contributed to this work, with special acknowledgement to Ding Wen Bao, Kun Cai, Qi Cai, Meaghan Coyle, Yunzhen He, Weixin Huang, Xiaodong Huang, James Kirby, Yaping Lai, Ting-Uei Lee, Yu Li, Zhi Li, Xiaoshan Lin, Hongjia Lu, Jiaming Ma, Xianchuan Meng, Yi Rong, Wei Shen, Vahid Shobeiri, Peng Wei, Yulin Xiong, Tao Xu, Xin Yan, Kai Yang, Yuan Yao, Feng Yuan, Zi-Long Zhao, Shiwei Zhou, Qiang Zhou, Yiyi Zhou, and Zicheng Zhuang.

I am deeply grateful to the Australian Research Council and RMIT University for their generous financial support through an Australian Laureate Fellowship (2020–2024). This funding enabled me to dedicate five years to full-time research and build a large team of talented researchers. This book represents part of the research outcomes from the project, titled *New Technologies for Delivering Sustainable Freeform Architecture.*

Melbourne, Australia Yi Min Xie
December 2024

Competing Interests The author has no competing interests to declare that are relevant to the content of this manuscript.

Contents

1	**Introduction**	1
1.1	Common Assumptions in Conventional Topology Optimization	1
1.2	Re-Examining Common Assumptions in Topology Optimization	2
	1.2.1 Pursuing a Unique, Globally Optimal Solution—The 'Best' Design	2
	1.2.2 Within a Specified Design Domain	3
	1.2.3 Under Prescribed Support Conditions	3
	1.2.4 For Predetermined Load Conditions	3
	1.2.5 Without Considering Designer's Aesthetic Preferences	4
1.3	Generalizing Topology Optimization by Eliminating Common Assumptions	4
1.4	Overview of the Book	5
	References	5
2	**Achieving Diverse and Competitive Designs**	7
2.1	Introduction	7
2.2	Changing Parameters in the Optimization Algorithm	8
2.3	Varying Optimization Parameters in Different Locations and Directions	9
	2.3.1 Varying the Filter Radius Across the Design Domain	9
	2.3.2 Assigning Local Volume Fractions to Subdomains	10
	2.3.3 Setting Distinct Filter Radii in Different Directions	14
2.4	Penalizing Parts of the Design Domain	17
2.5	Penalizing Existing Designs	18
	2.5.1 Penalizing the Initial Optimized Design	18
	2.5.2 Penalizing All Precedent Designs	19
2.6	Introducing Randomness into the Optimization Process	21
	2.6.1 Creating Random Voids in the Initial Model	21

	2.6.2 Penalizing Element Sensitivities by Random Coefficients	22
	2.6.3 Perturbing Load and Support Conditions	24
2.7	Explicitly Controlling the Structural Complexity	30
2.8	Conclusion	34
	References	34

3 Redefining the Design Domain ... 37
3.1	Introduction	37
3.2	Exploring Alternative Design Domains	37
3.3	Fixing Part of the Design Space	39
3.4	Setting Part of the Design Space as a Prohibited Region	41
3.5	Embedding a Geometric Pattern in the Design Domain	42
3.6	Selecting an Adequate Design Domain	44
3.7	Using an Adaptive Design Domain	48
3.8	Introducing Gaps Within the Design Domain	53
3.9	Conclusion	55
	References	56

4 Optimizing Support Locations ... 59
4.1	Introduction	59
4.2	Optimizing Support Locations and Structural Topology	61
	4.2.1 Problem Definition and Statement	61
	4.2.2 Optimization of Support Locations	63
	4.2.3 BESO Method for Optimizing Structural Topology	64
	4.2.4 Convergence Criterion	66
	4.2.5 Computational Workflow and Implementation	66
	4.2.6 Results of a Test Example	67
	4.2.7 Verification of Optimization Result	68
4.3	Discussion	69
	4.3.1 Material Model of Support Elements	69
	4.3.2 Effect of Mesh Density	71
	4.3.3 Effect of Support Stiffness	71
	4.3.4 Effect of Support Cost	74
	4.3.5 Alternative Optimization Methods	74
4.4	Applications	76
	4.4.1 Optimizing Support Locations of a 3D Shell Structure	76
	4.4.2 Designing a 3D Hinge Frame	77
4.5	Conclusion	78
	References	79

5 Optimizing Load Distributions ... 81
5.1	Introduction	81
5.2	Optimizing Load Locations	82
	5.2.1 Problem Definition and Statement	82
	5.2.2 Sensitivity Analysis	83

		5.2.3	Optimization Method	83
		5.2.4	Optimization Procedure	84
		5.2.5	Numerical Examples	85
	5.3	Finding Globally Optimal Locations of Multiple Point Loads Using a Single FEA	
				88
		5.3.1	Condensed Flexibility Matrix	89
		5.3.2	Possible Load Combinations or Permutations	90
		5.3.3	Numerical Examples	90
	5.4	Redistributing Load Magnitudes	93
		5.4.1	Modified Optimality Criteria Method	93
		5.4.2	Sequential Least Squares Quadratic Programming Method	95
		5.4.3	Interior-Point Method	96
		5.4.4	Numerical Examples	96
	5.5	Optimizing Load Directions	101
		5.5.1	Optimization Method	101
		5.5.2	Numerical Examples	101
	5.6	Conclusion	104
	References		105
6	**Human–Computer Interaction**			107
	6.1	Introduction	107
	6.2	Indicating Preferences by Drawing a Pattern	108
	6.3	Assigning Scores to Intermediate Designs	112
	6.4	Combining Drawing and Scoring in a Multi-solution System	113
	6.5	Performing Interactive Topology Optimization in Virtual Reality	118
	6.6	Improving Control of Subjective Preferences	123
	6.7	Conclusion	129
	References		130
7	**Practical Applications**			133
	7.1	Introduction	133
	7.2	Design of a Long-Span Steel–Concrete Composite Bridge	135
	7.3	Design of an Innovative Chair	139
	7.4	Conclusion	143
	References		144

Author Index 147

Index 151

About the Author

Yi Min Xie has been Distinguished Professor at Royal Melbourne Institute of Technology (RMIT), Laureate Fellow of the Australian Research Council, and Fellow of the Australian Academy of Technological Sciences and Engineering. His research focuses on computational design and digital fabrication of efficient and innovative structures. He has been instrumental in the original development and practical application of the widely used evolutionary structural optimization (ESO) and bi-directional evolutionary structural optimization (BESO) methods.

Abbreviations

BESO	Bi-directional evolutionary structural optimization
CAD	Computer-aided design
ESO	Evolutionary structural optimization
FEA	Finite element analysis
FFF	Fused filament fabrication
INP	Interior-point
INT	Integer
LVF	Local volume fraction
MOC	Modified optimality criteria
MR	Mixed reality
NC	Normalized compliance
OC	Optimality criteria
SIMP	Solid isotropic material with penalization
SLSQP	Sequential least squares quadratic programming
SP	Subjective preference
UDL	Uniformly distributed load
VF	Volume fraction
VR	Virtual reality

Chapter 1
Introduction

This chapter examines common assumptions widely adopted in conventional topology optimization. These assumptions include pursuing a unique, globally optimal solution—the 'best' design; conducting optimization within a predefined design domain, under prescribed load conditions and predetermined support conditions; and focusing on structural performance without considering aesthetic preferences of the designer or client. This book aims to demonstrate that such assumptions are not only unnecessary but can significantly limit the freedom and creativity in structural design.

1.1 Common Assumptions in Conventional Topology Optimization

There have been remarkable advances in topology optimization techniques over the past three decades since the seminal work by Bendsøe and Kikuchi (1988). These advancements have enabled a wide range of practical applications in aerospace, automotive, and other weight-conscious industries (Sigmund 2020). Moreover, topology optimization has been increasingly employed in architectural design, particularly for form-finding purposes (Beghini et al. 2014).

Following the well-established conventions in structural design, topology optimization adopts a set of common assumptions regarding its design objective and constraints. These assumptions are clearly reflected in the classic definition of topology optimization provided by Bendsøe and Sigmund (2004, p. 1):

> *The purpose of topology optimization is to find the optimal lay-out of a structure within a specified region. The only known quantities in the problem are the applied loads, the possible support conditions, the volume of the structure to be constructed and possibly some additional design restrictions such as the location and size of prescribed holes or solid areas.*

Similar statements have appeared repeatedly in numerous other publications on topology optimization, further reinforcing these assumptions as gold standards. However, throughout this book, I will challenge the necessity of these assumptions and demonstrate the benefits of removing them. Before delving into detailed discussions in the subsequent chapters, a brief re-evaluation of each assumption is presented below.

1.2 Re-Examining Common Assumptions in Topology Optimization

1.2.1 Pursuing a Unique, Globally Optimal Solution—The 'Best' Design

Bendsøe and Sigmund (2004) state that the purpose of topology optimization is to find the *optimal* layout of a structure. The term 'optimal' implies that one should aim to achieve the 'best' solution, a belief strongly held by many scholars in the field. During the early stages of my involvement in developing the evolutionary structural optimization (ESO) and bi-directional evolutionary structural optimization (BESO) methods (Xie and Steven 1997; Huang and Xie 2010b), I was often questioned by other researchers about whether we could guarantee our algorithms would produce a unique, globally optimal solution.

In mathematics, the uniqueness and optimality of a solution are indeed crucial considerations. However, in practical applications, achieving a unique, globally optimal design is often unnecessary. In fact, not only ESO/BESO but also other numerical methods for topology optimization generally do not yield a unique, globally optimal solution (Huang and Xie 2010a).

In recent years, instead of focusing on finding the 'best' solution, my team has been pursuing a nearly 'opposite' goal—developing algorithms to generate multiple solutions that exhibit distinctly different geometrical forms while maintaining acceptably high structural performance. These solutions are referred to as 'diverse and competitive designs'. This shift in mindset is a direct result of my active involvement in real-world projects (see Chap. 7) and my frequent dialogue with practising architects and engineers over the past decade. When working on practical projects, I have been repeatedly reminded that a unique, globally optimal solution, focused solely on structural performance, is usually of little value, as it often fails to meet other essential requirements, such as the aesthetic preferences of the designer or client. In this context, I quote the renowned Japanese structural designer Yoshikatsu Tsuboi (Tsuboi et al. 2012, p. 208):

> *Structure's beauty lies in the vicinity of its rationality.* (構造の美しさは合理の近傍にある.)

1.2.2 Within a Specified Design Domain

As Bendsøe and Sigmund (2004) suggest, topology optimization aims to determine the optimal layout of a structure *within a specified region*. This region is often referred to as the design space or design domain. Most examples in the literature often use a rectangular design domain for 2D problems and a cuboid for 3D problems, rather than strategically defining a design domain tailored to a specific problem.

In this book, we demonstrate that exploring alternative design domains can unlock numerous opportunities for design innovation, and selecting an appropriate design domain can significantly enhance structural performance without increasing the weight. In fact, the design domain itself can be considered a design variable, allowing it to adapt automatically according to structural or biomechanical requirements. We may even introduce some 'gaps' within the design domain to create unusual structures and metamaterials.

1.2.3 Under Prescribed Support Conditions

Traditionally, *support conditions* for a topology optimization problem are assumed to be *known quantities* (Bendsøe and Sigmund 2004). However, in practical projects, the number of supports and their locations are often not predetermined and can be adjusted. By treating the support conditions as additional design variables, we can significantly expand the design freedom, enabling the creation of more innovative and efficient structural designs compared to arbitrarily prescribing the support conditions. In particular, varying the number of supports and optimizing the support locations open new possibilities for architectural design.

1.2.4 For Predetermined Load Conditions

According to Bendsøe and Sigmund (2004), *applied loads* are also considered *known quantities* in topology optimization. In most cases found in the literature, the locations and directions of loads are predetermined before performing topology optimization. However, structural performance is often extremely sensitive to applied loads. Even a minor change in the load distributions—including the load locations, directions, and magnitudes—can significantly influence the structural response.

It is therefore highly desirable to apply the loads at the 'sweet spots'—the optimal locations—so that the structure can perform at its best (Cross 1998). Conversely, identifying the most dangerous load distribution is crucial for ensuring the safety of certain structures like sports stadiums, which may occasionally experience excessive and unpredictable crowd movements.

Since 2022, my team has been developing optimization algorithms to identify the 'best' and 'worst' load distributions, corresponding to the highest and lowest structural performance, respectively. Further, these optimized load distributions can be incorporated into the iterative process of topology optimization to achieve the safest or most efficient structural design.

1.2.5 Without Considering Designer's Aesthetic Preferences

Topology optimization, at its core, is a mathematical approach primarily used in engineering design. Beyond engineering, it has also found applications in fields such as architectural design (Cui et al. 2003; Beghini et al. 2014; Morales-Beltran and Mostafavi 2022) and furniture design (Ma et al. 2021). In these fields, aesthetics and other subjective preferences of the designer or client often play a much more significant role than in engineering.

A key challenge in applying topology optimization to architectural or furniture design lies in the difficulty in quantifying or codifying subjective preferences mathematically. Furthermore, while a designer may initially have a vague or tentative idea for the conceptual design of a project, most of the details tend to be fluid and evolving throughout the entire design exploration process.

As seen from Bendsøe and Sigmund (2004) and nearly all the literature in the field, subjective preferences are rarely considered as part of topology optimization. Occasionally, aesthetics is addressed by post-processing the results of topology optimization. How can subjective preferences be incorporated directly into the topology optimization process? How can the designer interact with the computer to influence or control the final design outcome? These are some of the questions my team has been exploring in recent years as we work towards establishing a human–computer interaction platform aimed at creating innovative designs that balance structural performance and subjective preferences.

1.3 Generalizing Topology Optimization by Eliminating Common Assumptions

This book advocates for a generalized topology optimization approach that eliminates all the common assumptions discussed in the previous section. We introduce techniques and algorithms to systematically remove each of these assumptions. Through numerous examples, we demonstrate that these assumptions are not only unnecessary but also impose severe limitations on design freedom and hinder creativity in structural design.

The core concepts of the generalized topology optimization approach were first outlined by Xie (2022) in the context of architectural design. Since then, we have significantly expanded these concepts and related techniques, broadening their applicability to a much wider range of structural design problems.

1.4 Overview of the Book

This book provides a systematic discussion of the limitations of each of the common assumptions in conventional topology optimization. It presents techniques for eliminating each assumption and demonstrates the benefits of doing so.

Chapter 2 explores a variety of techniques for generating diverse and competitive designs that are geometrically distinct yet structurally efficient.

Chapter 3 considers alternative design domains to achieve innovative and efficient designs, and to create unusual structures.

Chapter 4 introduces a method for the simultaneous optimization of support locations and structural topology.

Chapter 5 presents methods for optimizing load locations and directions, as well as redistributing load magnitudes.

Chapter 6 discusses a human–computer interaction platform based on topology optimization, integrating subjective preferences and structural performance.

Chapter 7 provides examples of practical projects that have utilized and benefited from some of the concepts and techniques of the generalized topology optimization approach.

References

Beghini, L. L., Beghini, A., Katz, N., Baker, W. F. and Paulino, G. H. (2014) Connecting architecture and engineering through structural topology optimization. *Eng. Struct.* **59**, 716–726.

Bendsøe, M. P. and Kikuchi, N. (1988) Generating optimal topologies in structural design using a homogenization method. *Comput. Methods Appl. Mech. Eng.* **71**, 197–224.

Bendsøe, M. P. and Sigmund, O. (2004) *Topology Optimization: Theory, Methods and Applications.* 2nd edition. Berlin: Springer.

Cross, R. (1998) The sweet spot of a baseball bat. *Am. J. Phys.* **66**, 772–779.

Cui, C., Ohmori, H. and Sasaki, M. (2003) Computational morphogenesis of 3D structures by extended ESO method. *J. Int. Assoc. Shell Spat. Struct.* **44**, 51–61.

Huang, X. and Xie, Y. M. (2010a) A further review of ESO type methods for topology optimization. *Struct. Multidisc. Optim.* **41**, 671–683.

Huang, X. and Xie, Y. M. (2010b) *Evolutionary Topology Optimization of Continuum Structures: Methods and Applications.* Chichester: John Wiley & Sons.

Ma, J., Li, Z., Zhao, Z.-L. and Xie, Y. M. (2021) Creating novel furniture through topology optimization and advanced manufacturing. *Rapid Prototyp. J.* **27**, 1749–1758.

Morales-Beltran, M. and Mostafavi, S. (2022) Topology optimization in architectural design. *Proc. 40th Educ. Res. Comput. Aided Archit. Des. Eur. (eCAADe)*, Ghent, 13–16 September 2022, 589–598.

Sigmund, O. (2020) EML webinar overview: Topology optimization—status and perspectives. *Extreme Mech. Lett.* **39**, 100855.

Tsuboi, Y., Kawaguchi, M., Sasaki, M., Ohsaki, M., Ueki, T., Takeuchi, T., Kawabata, M., Kawaguchi, K. and Kanebako Y. (2012) Mechanics, material and structural design. *Archit. Technol.* (in Japanese).

Xie, Y. M. (2022) Generalized topology optimization for architectural design. *Archit. Intell.* **1**, 2.

Xie, Y. M. and Steven, G. P. (1997) *Evolutionary Structural Optimization.* London: Springer.

Open Access This chapter is licensed under the terms of the Creative Commons Attribution 4.0 International License (http://creativecommons.org/licenses/by/4.0/), which permits use, sharing, adaptation, distribution and reproduction in any medium or format, as long as you give appropriate credit to the original author(s) and the source, provide a link to the Creative Commons license and indicate if changes were made.

The images or other third party material in this chapter are included in the chapter's Creative Commons license, unless indicated otherwise in a credit line to the material. If material is not included in the chapter's Creative Commons license and your intended use is not permitted by statutory regulation or exceeds the permitted use, you will need to obtain permission directly from the copyright holder.

Chapter 2
Achieving Diverse and Competitive Designs

To enhance the practical applicability of topology optimization results in engineering and architecture, it is highly desirable to generate multiple solutions to a given problem that are not only structurally efficient but also geometrically distinct. This chapter presents various techniques we have developed in recent years to achieve diverse and competitive designs *without altering actual load and support conditions*. These techniques include changing optimization parameters within the optimization algorithm or across the design domain, penalizing existing designs or specific regions of the design domain, introducing randomness into the optimization process, and explicitly controlling the structural complexity.

2.1 Introduction

In the long tradition of structural optimization, local optima are often frowned upon by researchers. Many mathematicians and some engineers are obsessed with attaining the 'unique' and 'globally optimal' solution from optimization algorithms. However, I find that most architects I have collaborated with are less inclined to insist on or accept the 'optimal solution', because it often deviates from their artistic intuitions and fails to meet various functional requirements. The 'best design' based solely on structural performance is usually of low value in architectural practice. During the early stages of a design project, an engineer who overzealously extols the virtues of such a unique solution without considering alternative options may inadvertently cause the breakdown of a nascent collaboration with an architect or a client.

To address this bottleneck, my team has developed a series of techniques since 2019 aimed at producing multiple designs that have distinctly different configurations but possess similar structural performance to that of the optimal solution (e.g., Xie et al. 2019; Yang et al. 2019; He et al. 2020, 2022, 2023; Zhao et al. 2020; Cai et al. 2021; Xie 2022; Yan et al. 2022, 2023; Xiong et al. 2024, 2025). We demonstrate that vastly different designs can be obtained by sacrificing only a small amount (e.g., 5%) of the structural performance, and in some cases even improving it. Such *diverse* and *competitive* designs offer architects and clients a broad range of design options while maintaining a high level of structural efficiency. In the following sections, we present a variety of techniques for achieving diverse and competitive designs without altering actual load and support conditions.

2.2 Changing Parameters in the Optimization Algorithm

The easiest way to obtain diverse and competitive designs is to change one or more input parameters in the optimization algorithm. For instance, one of the commonly used parameters in various topology optimization methods is the filter radius r_{min}, which controls, indirectly, the minimum member size in the optimized structure (Sigmund and Petersson 1998; Huang and Xie 2007). Figure 2.1 illustrates the results of generating multiple designs for a short cantilever using the bi-directional evolutionary structural optimization (BESO) method (Huang and Xie 2010) by simply varying the filter radius r_{min} from 4.5 to 1.0. The cantilever is fixed on the left-hand side and is subjected to a vertical load at the midpoint of the right-hand side.

We obtain four distinctly different designs through this approach. Using a larger r_{min} results in fewer members but larger sizes (see Fig. 2.1b). In contrast, a smaller r_{min} leads to more members but smaller sizes (see Fig. 2.1e). For the same amount of material, these four geometrically distinct designs exhibit less than 1.4% difference in their structural compliance (the inverse of the overall stiffness).

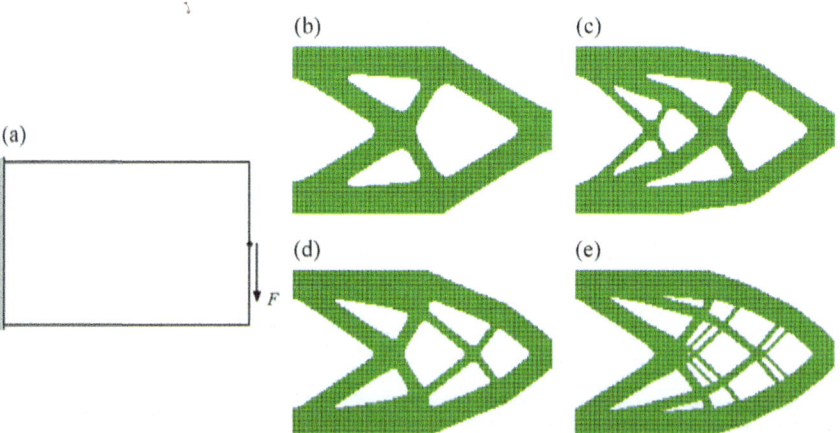

Fig. 2.1 Multiple designs of a short cantilever obtained by changing the filter radius r_{min}: **a** load and support conditions; **b** $r_{min} = 4.5$; **c** $r_{min} = 3.0$; **d** $r_{min} = 2.0$; **e** $r_{min} = 1.0$

This straightforward technique of changing input parameters within optimization algorithms is highly effective in generating diverse and competitive designs, offering a simple yet powerful way to explore a wide range of design possibilities.

2.3 Varying Optimization Parameters in Different Locations and Directions

2.3.1 Varying the Filter Radius Across the Design Domain

In the previous section, topology optimization is performed with a constant filter radius throughout the entire design domain. However, by assigning distinct filter radii to different regions or varying the filter radius across the design domain, a broader range of diverse and competitive solutions can be achieved. As noted earlier, the filter radius influences the minimum member size in the optimized structure. When a large filter radius is assigned to a region, thick members tend to emerge locally. Conversely, a small filter radius leads to slender members. Typically, the minimum member size in a region is approximately twice the local filter radius, if the volume fraction (VF) is small enough.

To implement this approach, the design domain can be divided into a finite number of regions, with each region assigned a distinct filter radius. Alternatively, the filter radius can be set to vary gradually across the design domain, thereby creating a smooth transition of minimum member sizes across different regions.

We use one example to demonstrate the effect of varying the filter radius across the design domain. Figure 2.2 shows a spherical shell dome that is 19 m high, with a base radius of 32 m and an oculus of 4 m in radius at the top (Yan et al. 2022). The shell has a thickness of 0.3 m and is subjected to a gravity load in the vertical direction and a torsional load along the edge of the oculus (indicated by the blue arrow in Fig. 2.2a). Fixed supports are assigned to 18 equally spaced areas at the base of the dome and a thin layer along the oculus boundary is set as non-design domain (see Fig. 2.2a). This example is inspired by the iconic Palazzetto dello Sport (Small Sport Palace) in Rome, engineered by Pier Luigi (Nervi 2018).

By applying either a constant or variable filter radius across the design domain, we obtain six geometrically distinct designs. Figure 2.3a, b shows the solutions from using uniform filter radii of 100 mm and 500 mm, respectively, resulting in nearly equal member sizes across the entire design domain. Next, we vary the filter radius linearly from the outer boundary to the centre of the dome, either from 100 to 500 mm (see Fig. 2.3c) or from 500 to 100 mm (see Fig. 2.3e). Additionally, we apply a bi-linear variation of the filter radius from 100 to 500 mm and back to 100 mm (see Fig. 2.3d) or from 500 to 100 mm and back to 500 mm (see Fig. 2.3f). It can be seen that the member sizes change gradually and smoothly across different regions in accordance with the local filter radius values.

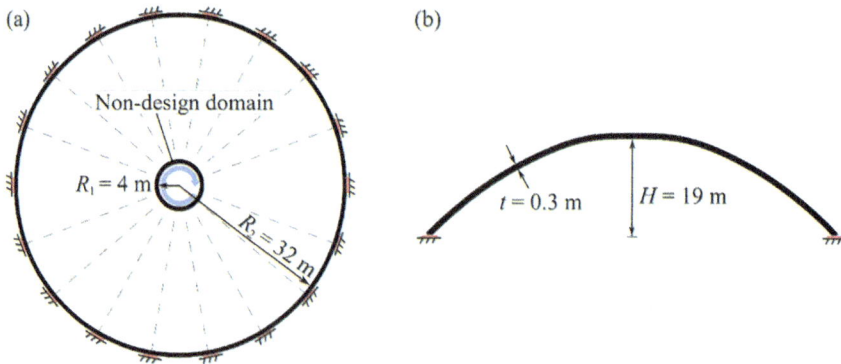

Fig. 2.2 Boundary and load conditions of a spherical dome: **a** top view; **b** side view (reprinted from Yan et al. 2022, with permission from authors)

To evaluate the structural performance of each design, we calculate the normalized compliance (NC), defined as the ratio of a structure's compliance to that of a reference design. In this case, the solution with a constant filter radius of 100 mm, shown in Fig. 2.3a, is chosen as the reference design. The NC values of the four designs with variable filter radius across the design domain (see Fig. 2.3c–f) range from 0.999 to 1.075, indicating that the maximum loss in structural performance corresponds to a 7.5% increase in compliance. This level of performance loss is relatively small, considering that the designs have changed from having nearly uniform member sizes to ones with varying member sizes across a wide range. From these observations, we can conclude that the solutions in Fig. 2.3 are not only geometrically diverse but also structurally competitive.

2.3.2 Assigning Local Volume Fractions to Subdomains

Traditionally, topology optimization is performed with a constraint on the total structural volume. Typically, the volume constraint is given in terms of a volume fraction, defined as the ratio of the volume of the structure to that of the maximum allowable space. Although such a global constraint on the volume usually results in a structurally efficient design, it may not always meet the designer's expectations. This is because the material may become overly concentrated in certain areas, leaving other regions nearly empty, as shown in Fig. 2.4 for the design of a high-rise building facade (Yan et al. 2023).

In this example, the building is 20 m wide and 100 m tall. It is fixed at the base and subjected to two separate load cases: one with the wind acting on the left-hand side of the building and the other on the right-hand side (see Fig. 2.4a). With a uniform filter radius of 0.6 m and a uniform volume fraction of 50%, the BESO method produces the conventional result shown in Fig. 2.4b. However, this design may not be ideal

2.3 Varying Optimization Parameters in Different Locations and Directions

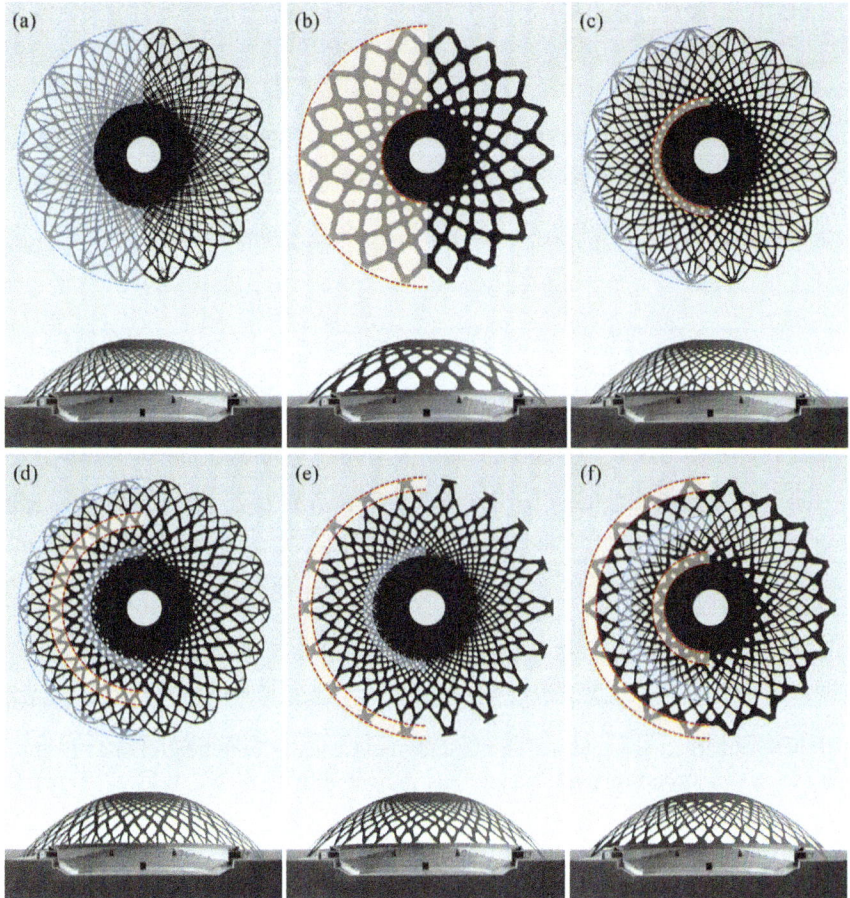

Fig. 2.3 Optimized shell dome designs with a constant or variable filter radius r_{min}: **a** r_{min} = 100 mm, NC = 1; **b** r_{min} = 500 mm, NC = 1.180; **c** r_{min} = 100 → 500 mm, NC = 0.999; **d** r_{min} = 100 → 500 → 100 mm, NC = 1.014; **e** r_{min} = 500 → 100 mm, NC = 1.053; **f** r_{min} = 500 → 100 → 500 mm, NC = 1.075 (reprinted from Yan et al. 2022, with permission from authors)

for a building facade, as the extensive solid areas on the lower levels would obstruct daylight and the large empty spaces on the upper levels would fail to meet other structural and functional requirements.

To address the issue of unsatisfactory material distribution, we divide the building facade into 10 subdomains and assign a different local volume fraction (LVF) to each subdomain (see Fig. 2.5). For instance, in Case 1, the LVFs are set to 10% and 90% for the bottom and top subdomains, respectively, while the LVFs of intermediate subdomains are generated by a linear interpolation between 10 and 90%. Using the same filter radius of 0.6 m but different LVF distributions, we obtain nine geometrically distinct designs, as shown in Fig. 2.5. Although the nine cases have different LVF

Fig. 2.4 Design of a high-rise building facade: **a** boundary and load conditions; **b** conventional BESO result (reprinted from Yan et al. 2023, with permission from Elsevier)

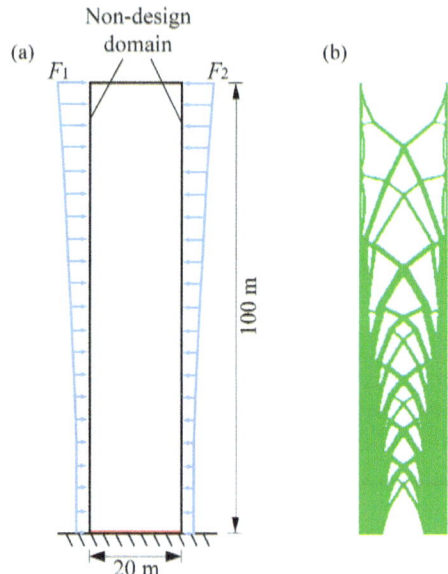

variations, their total volume fractions are all equal to 50%. The structural performance of each design is measured by the normalized its compliance with respect to the conventional BESO design shown in Fig. 2.4b.

It is reasonable to reject the designs shown in Cases 1–4 on the grounds of significant losses in structural performance (as indicated by their NC values). In particular, Cases 1 and 2 feature excessively heavy upper levels, rendering these designs structurally irrational.

Nonetheless, the solutions shown in Cases 5–9 can be regarded as diverse and competitive designs, as their compliance values increase by no more than 17.1% compared to the conventional BESO result. In particular, the designs in Cases 6 and 7 have substantially alleviated the problems of excessive solid areas on the lower levels and nearly empty spaces on the upper levels, while their compliance values only increase by 9.0% and 3.3%, respectively.

To further demonstrate potential practical applications of varying the volume fraction across the design domain, we consider another example (Yan et al. 2023). A shell dome, similar to the one shown in Fig. 2.2 but with different dimensions, is used here: the oculus radius $R_1 = 2.5$ m, the base radius $R_2 = 19$ m, the dome height $H = 5$ m, and the shell thickness $t = 0.3$ m. The boundary and load conditions are also similar to those illustrated in Fig. 2.2a. With a uniform filter radius of 0.3 m and a uniform volume fraction of 50%, the BESO method produces the conventional result shown in Fig. 2.6a. This geometrical pattern looks interesting, but it presents a significant issue—the large solid area near the centre is not suitable for the roof of a sports arena, as it blocks daylight and appears excessively heavy.

Note that the large solid area near the centre is not a result of it being designated as a non-design domain. In fact, only a very thin layer along the oculus boundary

2.3 Varying Optimization Parameters in Different Locations and Directions

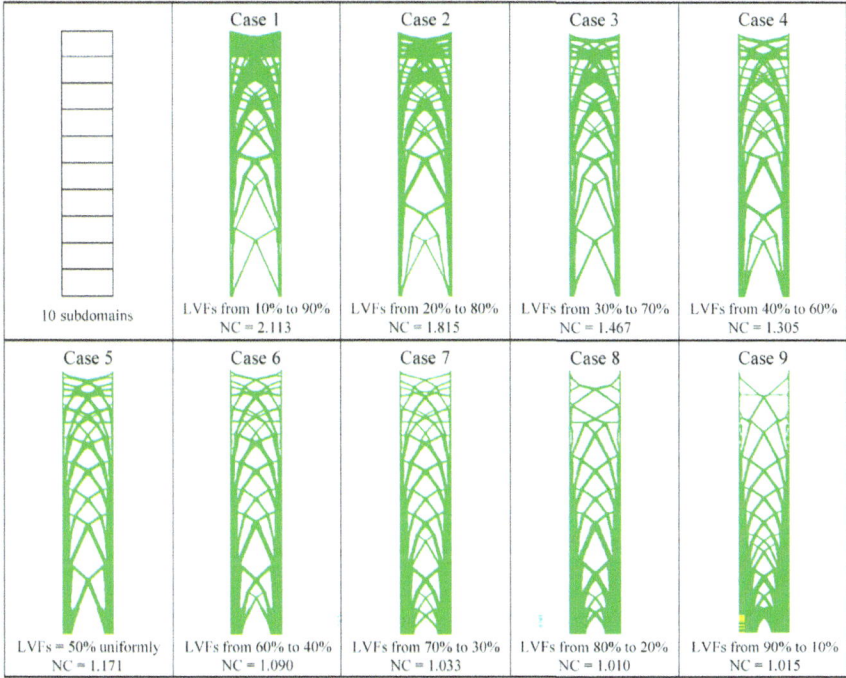

Fig. 2.5 Diverse designs for a high-rise building facade from assigning different LVFs to subdomains (reprinted from Yan et al. 2023, with permission from Elsevier)

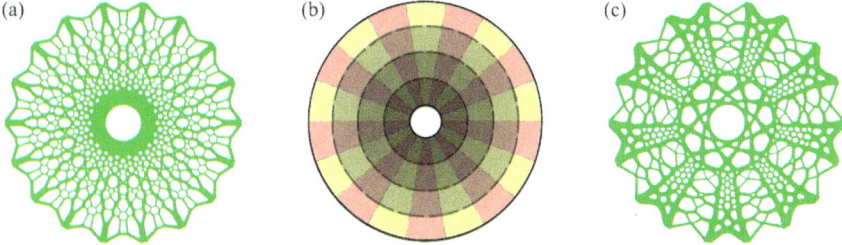

Fig. 2.6 Distinct designs for a shell dome from assigning uniform and variable volume fractions: **a** conventional BESO result with a uniform volume fraction; **b** subdomains for assigning different LVFs; **c** alternative design with different LVFs for subdomains, NC = 1.013 (reprinted from Yan et al. 2023, with permission from Elsevier)

is set as the non-design domain (see Figs. 2.2a and 2.6c). When a uniform volume fraction is applied to the entire design domain, a thick solid annulus consistently forms around the oculus, as observed in all seven optimized designs in Figs. 2.3 and 2.6a. These results seem to indicate that the thick solid annulus is essential for maintaining the structural performance. But is it really so?

To answer this question, we divide the dome into four concentric annuli and then subdivide each annulus equally into 18 subdomains (see Fig. 2.6b). Each subdomain can be assigned a different LVF, with a lower LVF leading to less material being distributed to this subdomain. If a designer wishes to create holes or cavities in a specific region, it can be achieved by simply setting the LVF for the corresponding subdomain to less than 1.

Among numerous possibilities, we set the LVF of each yellow and red subdomain to 33.33% and 67.67%, respectively, resulting in an overall volume fraction of 50% for each annulus (and for the whole structure). This leads to an alternative dome design shown in Fig. 2.6c, with more material allocated to the red regions and less material to the yellow ones. Notably, the large solid area in Fig. 2.6a has been transformed into a perforated, elegant gridshell in Fig. 2.6c. Although the geometric patterns of the two designs in Fig. 2.6a, c are completely different, their compliance values differ by only 1.3%. This outcome is truly remarkable!

2.3.3 Setting Distinct Filter Radii in Different Directions

In traditional topology optimization methods, the filter radius r_{min} is treated as a scalar, meaning a single value is assigned to each element or node. For 3D problems, r_{min} represents the radius of a sphere within which the filtering is applied (see Sect. 4.2.3 for details). To increase the diversity of topology optimization results, an anisotropic filter can be employed, where the filter radius is represented by a vector $\mathbf{r}$, with different values in x, y, and z directions, i.e., $\mathbf{r} = [r_x, r_y, r_x]^T$. Instead of a radius of a sphere, r_x, r_y, and r_z can be interpreted as the half-lengths of the principal axes of an ellipsoid, oriented such that its principal axes align with the x, y, and z directions.

An anisotropic filter can be conveniently implemented using an alternative filtering method based on Helmholtz-type differential equations (Lazarov and Sigmund 2011; Wang et al. 2020; Xiong et al. 2025). Rather than delving into the theoretical formulations of the anisotropic filter, here we focus our discussion on how to achieve diverse and competitive designs by setting distinct filter radii in different directions.

Figure 2.7a illustrates a 3D cantilever with dimensions of 30 mm, 10 mm, and 10 mm along the x, y, and z directions, respectively. A non-design bar, highlighted in yellow and measuring 0.5 mm in both width and height, is used to apply a uniformly distributed load. When an isotropic filter with $r_x = r_y = r_z = 3$ mm is applied and the volume fraction is set to 30%, we obtain a conventional topology optimized solution from the BESO method, as shown in Fig. 2.7b.

As mentioned previously, the filter radius influences the minimum member size in the optimized structure, with a larger filter radius leading to larger member sizes. When using the anisotropic filter, if the filter radius in a particular direction significantly exceeds the design domain's dimension in that direction, the resulting structure becomes extruded along that axis, as illustrated in Fig. 2.8a–c. Interestingly, the

2.3 Varying Optimization Parameters in Different Locations and Directions

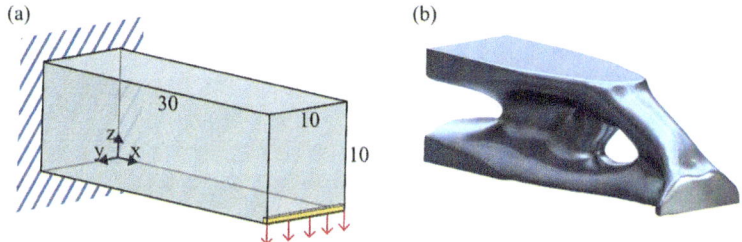

Fig. 2.7 Design optimization of a 3D cantilever: **a** boundary and load conditions; **b** conventional solution from the BESO method using an isotropic filter with $r_x = r_y = r_z = 3$ mm (reprinted from Xiong et al. 2025, licensed under CC-BY 4.0)

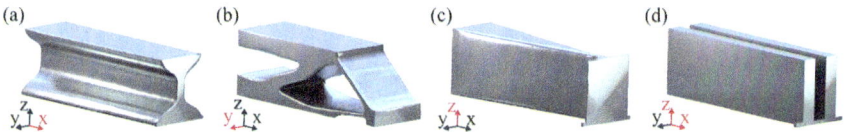

Fig. 2.8 Diverse designs for a 3D cantilever using anisotropic filtering: **a** $r_x = 100$ mm, $r_y = 3$ mm, $r_z = 3$ mm, NC = 1.106; **b** $r_x = 3$ mm, $r_y = 100$ mm, $r_z = 3$ mm, NC = 1.240; **c** $r_x = 3$ mm, $r_y = 3$ mm, $r_z = 100$ mm, NC = 1.214; **d** $r_x = 100$ mm, $r_y = 3$ mm, $r_z = 100$ mm, NC = 1.595 (reprinted from Xiong et al. 2025, licensed under CC-BY 4.0)

extruded design shown in Fig. 2.8a bears a striking resemblance to the commonly used I-shaped 'universal beam'.

It is noted that extrusion is a widely used manufacturing technique for producing metallic and polymeric materials and structures (Saha 2000; Volk et al. 2022). By applying the anisotropic filer, we can readily generate diverse designs that satisfy the manufacturing constraint associated with extrusion.

We can also use the anisotropic filter to create plate-like structures by assigning large filter radii in two directions and a small filter radius in the third direction, either to the entire design domain or to a specific region. This is equivalent to simultaneously extruding the structure in two directions. Figure 2.8d shows a design with 'plates' in the xz plane as a result of using large values for r_x and r_z, while keeping r_y relatively small. Note that the thickness of the plate depends on the prescribed volume fraction and the small filter radius.

Conventional topology optimization methods often produce truss-like structures, particularly when the volume fraction is relatively small. In contrast, plate-like structures generated using the anisotropic filter offer architects and engineers alternative geometrical features. Furthermore, compared to their truss-like counterparts, plate-like structures usually require less support material for additive manufacturing (Rieser and Zimmermann 2023), and may sometimes exhibit superior structural performance (Sigmund et al. 2016) as demonstrated by an example in Sect. 2.7 (see Fig. 2.26).

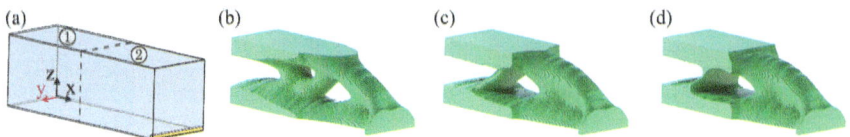

Fig. 2.9 Effect of varying the filter radius in Region ① in the y direction: **a** two regions of the design domain; **b** $r_{y1} = 5$ mm, NC $= 1.023$; **c** $r_{y1} = 10$ mm, NC $= 1.080$; **d** $r_{y1} = 15$ mm, NC $= 1.101$ (reprinted from Xiong et al. 2025, licensed under CC-BY 4.0)

In the caption of Fig. 2.8, we have provided the normalized compliance of each design with respect to the conventional solution shown in Fig. 2.7b. The extruded designs are geometrically distinct from the conventional design, and their compliance values have increased by 10.6%, 24.0%, 21.4%, and 59.5%, respectively. Nonetheless, for some practical applications, these extruded designs may be useful or even preferred when other requirements or constraints need to be satisfied.

To further enhance the geometric diversity, we divide the design space into two regions, as illustrated in Fig. 2.9a, with different filter radii assigned to the two regions. Figure 2.9b–d shows the results obtained from using $r_{y1} = 5$ mm, 10 mm, and 15 mm, respectively, while keeping $r_{x1} = r_{z1} = 3$ mm and $r_{x2} = r_{y2} = r_{z2} = 3$ mm. It is seen that as r_{y1} gradually increases, Region ① transforms into an extruded structure, while Region ② evolves into a more conventional free-form configuration. When $r_{y1} = 5$ mm, the geometrical transition across the two regions is relatively smooth (see Fig. 2.9b), in contrast to the results for $r_{y1} = 10$ mm and $r_{y1} = 15$ mm (see Fig. 2.9c, d). In terms of structural performance, the compliance values of the designs in Fig. 2.9b–d have increased by 2.3%, 8.0%, and 10.1%, respectively, compared to the conventional solution shown in Fig. 2.7b.

The next example is inspired by a real project my team was involved in, for the design of a library in Hangzhou, China. Figure 2.10 illustrates a significantly simplified version of the actual project. The design space for one quarter of the building is shown in Fig. 2.10a, featuring two symmetric planes measuring 10,000 mm × 12,000 mm and 7000 mm × 12,000 mm, respectively. The middle and top layers, highlighted in yellow, represent floors of the building and are designated as non-design domain.

When an isotropic filter with $r_x = r_y = r_z = 300$ mm is applied and the volume fraction is set to 15%, we obtain the solution shown in Fig. 2.10b, reflecting a typical outcome of conventional topology optimization. However, this design was rejected by the chief architect of the project. Further discussions with the architect revealed that, based on his aesthetic preferences and concerns about construction costs, he would prefer straight walls and columns, especially in the lower portion of the building.

To achieve structurally efficient designs with straight walls and columns, we assign a very large value to the filter radius in the vertical direction for either the entire building or only the lower portion (Region ②). This results in two completely different designs shown in Fig. 2.10c, d. When r_z or r_{z2} is set to 20,000 mm, which far exceeds the vertical dimension of the design domain, the structural components in

2.4 Penalizing Parts of the Design Domain

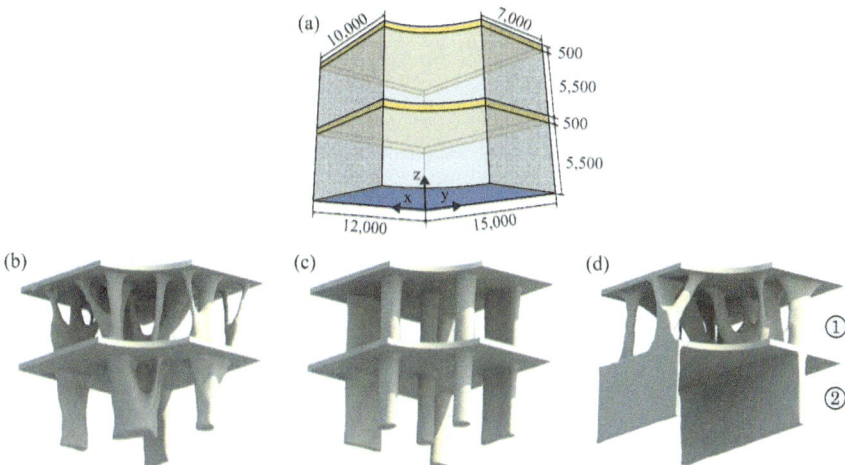

Fig. 2.10 Diverse designs for a library building: **a** design space of a quarter of the building (unit: mm); **b** $r_x = r_y = r_z = 300$ mm for the entire building, NC $= 1$; **c** $r_z = 20{,}000$ mm for the entire building, NC $= 1.155$; **d** $r_{z1} = 300$ mm for Region ①, $r_{z2} = 20{,}000$ mm for Region ②, NC $= 1.100$ (reprinted from Xiong et al. 2025, licensed under CC-BY 4.0)

the entire building or Region ② are forced to become extruded structures in the vertical direction, i.e., straight walls and columns, as preferred by the architect. Although the alternative solutions in Fig. 2.10c, d are geometrically distinct from the conventional optimization result in Fig. 2.10b, their compliance values have only increased by 15.5% and 10.0%, respectively. Therefore, these alternative solutions can be regarded as diverse and competitive designs.

2.4 Penalizing Parts of the Design Domain

Sometimes, a designer may want certain regions of a structure to appear lighter and more transparent, while making other areas sturdier and more solid. This can be achieved by artificially reducing the fitness of elements in specific parts of the structure. The fitness of an element can be determined by the so-called sensitivity number which approximately predicts the element's contribution to the objective function, such as the structural stiffness. Details about the element sensitivity number are provided in Sect. 4.2.3.

A simple technique for shifting material away from a certain part of the design domain is to penalize elements in this area by multiplying their sensitivity numbers by a constant coefficient less than 1 in every iteration of the optimization process. In doing so, these elements will have less chance to survive in the final design. This concept is illustrated using a cantilever example shown in Fig. 2.11. By applying the conventional BESO method, we obtain a symmetric design shown in Fig. 2.11b.

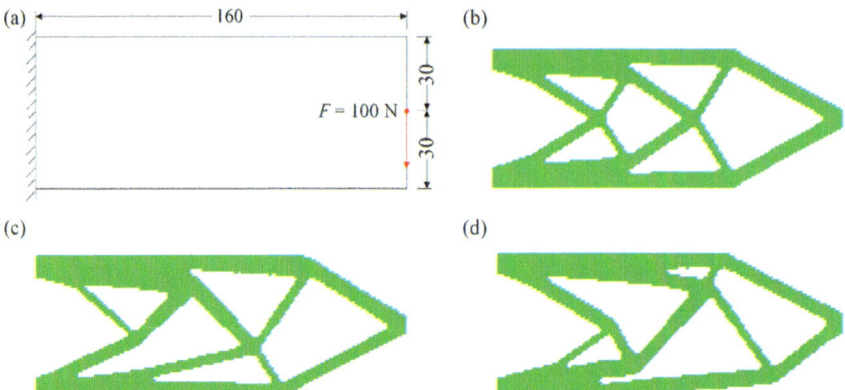

Fig. 2.11 Multiple designs for a cantilever: **a** boundary and load conditions; **b** initial optimized design without penalization; **c** and **d** alternative designs with penalty coefficients of 0.6 and 0.7 applied to the lower half, respectively (reprinted from Yang et al. 2019, with permission from Elsevier)

However, when the element sensitivity of the lower half of the design domain is penalized by constant coefficients of 0.6 and 0.7, respectively, two distinctly different designs emerge, as shown in Fig. 2.11c, d, with less material appearing in the lower half and more material in the upper half. Despite the significant geometric differences among the three designs, the compliance values of the two alternative solutions in Fig. 2.11c, d only increase by 2% and 4%, respectively, compared to that of the initial design in Fig. 2.11b.

2.5 Penalizing Existing Designs

We present two slightly different methods to create diverse and competitive designs based on penalizing certain parts that appear in previously optimized designs.

2.5.1 Penalizing the Initial Optimized Design

In the first method, we penalize a prescribed percentage of the least-efficient solid elements in the initial optimized design obtained by the conventional BESO process. By applying different penalty percentages, a series of new designs can be generated. We demonstrate this approach using a long-span structure example. The load and support conditions are shown in Fig. 2.12a. The initial optimized design is presented in Fig. 2.12b, and a subsequent design generated by penalizing the initial optimized design is given in Fig. 2.12c. In this case, the penalty percentage is set to 100% of

2.5 Penalizing Existing Designs

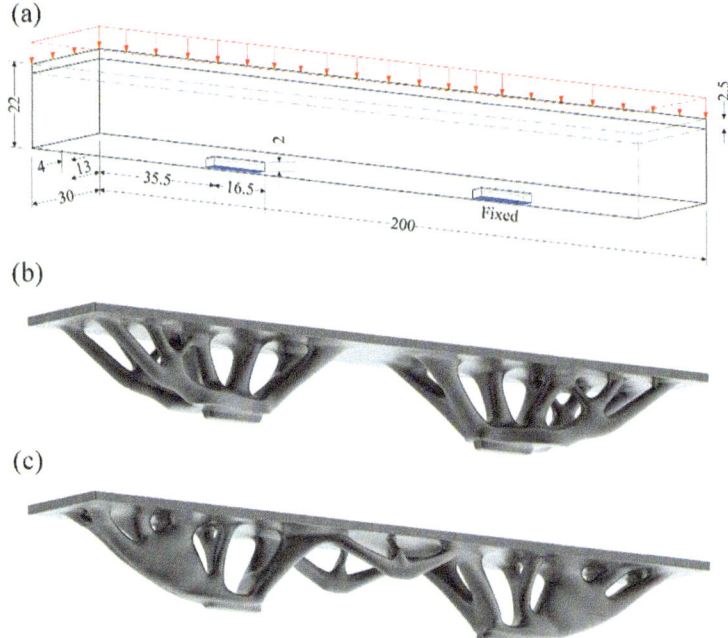

Fig. 2.12 Multiple designs for a long-span structure: **a** load and support conditions; **b** initial optimized design; **c** subsequent design generated by penalizing the initial optimized design (reprinted from Yang et al. 2019, with permission from Elsevier)

the remaining solid elements shown in Fig. 2.12b. For the same volume of material, the compliance of the subsequent design in Fig. 2.12c is 10% higher than that of the initial optimized design in Fig. 2.12b.

It is worth noting that a magnificent structure closely resembling the initial optimized design (see Fig. 2.12b) has been constructed at the Qatar National Convention Centre, designed by the renowned architect Arata Isozaki and his team using an extended evolutionary structural optimization (ESO) method (Cui et al. 2003; Sasaki 2005). The subsequent design in Fig. 2.12c could serve as an alternative structural form for a new project with similar or identical load and support conditions.

2.5.2 Penalizing All Precedent Designs

In the second method, *all* precedent designs are included in the penalization process when searching for subsequent structural forms. During each design cycle, a certain percentage of the least-efficient solid elements in every precedent design is penalized, making these elements less likely to reappear in the subsequent design.

Fig. 2.13 Multiple designs for a bridge structure: **a** load and support conditions; **b** initial optimized design; **c** and **d** subsequent designs generated by penalizing all precedent design(s) (reprinted from Yang et al. 2019, with permission from Elsevier)

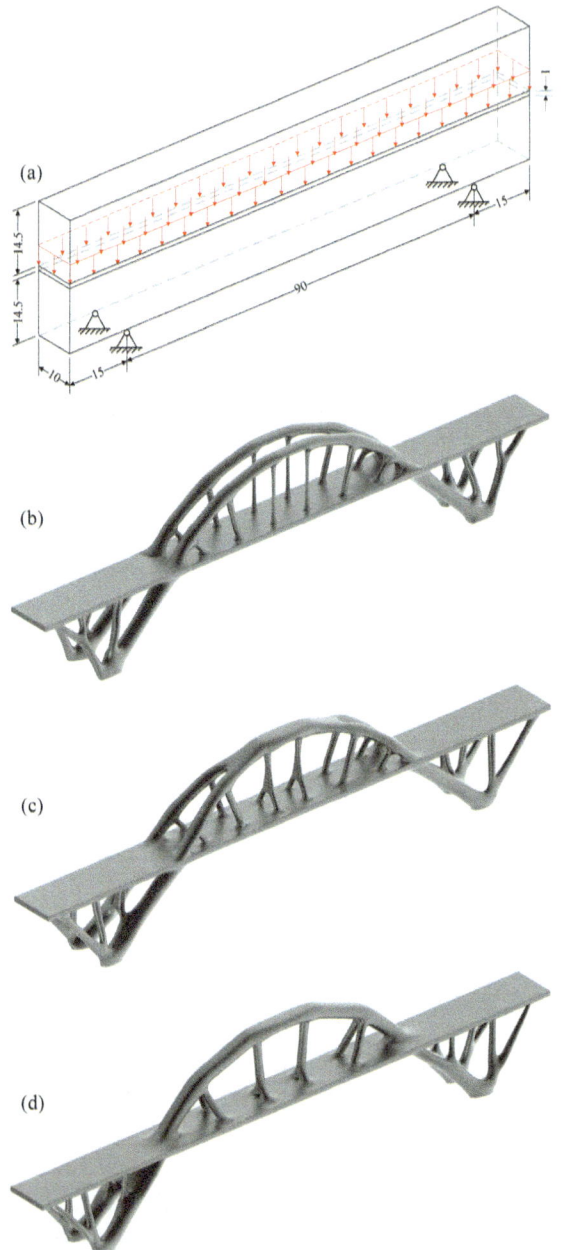

Figure 2.13 illustrates the application of the second method using a bridge structure example. The load and support conditions are given in Fig. 2.13a. The initial optimized design is presented in Fig. 2.13b. The first subsequent design (see Fig. 2.13c) is obtained by penalizing the only precedent form so far (see Fig. 2.13b), while the second subsequent design (see Fig. 2.13d) is obtained by penalizing both precedent forms (see Fig. 2.13b, c). The penalty percentage is set to 10% of the least-efficient solid elements remaining in each precedent design. The arches and other details of the three bridges in Fig. 2.13b–d differ substantially, yet the compliance values of the two subsequent designs are only 3% higher than that of the initial optimized design. All three designs are structurally efficient and aesthetically pleasing, with the final choice potentially influenced by other factors such as the arrangement of traffic lanes.

2.6 Introducing Randomness into the Optimization Process

A different technique for altering the design outcomes is to introduce some randomness into the optimization process. This technique can produce many unexpected results, allowing the designer to obtain a different structural form each time the optimization process is re-run. There are various ways to incorporate randomness into topology optimization models. Here, we present three techniques that we have developed in recent years (Xie et al. 2019; He et al. 2020; Xiong et al. 2024).

2.6.1 Creating Random Voids in the Initial Model

Figure 2.14a shows the boundary and load conditions of a cantilever. Its conventional optimized design using the BESO method is given in Fig. 2.14b. By introducing random voids in the initial finite element model (see Fig. 2.14c, e), we deliberately divert the optimization process away from its normal course, using the voids as 'noise'. Figure 2.14d, f shows two distinctly different designs generated after introducing random voids to the initial finite element model. It should be noted that some of these voids are converted into solid elements in later iterations at locations where the material is needed for structural efficiency. Compared to the conventional optimized design (see Fig. 2.14b), the compliance of the design in Fig. 2.14d is only 4% higher and, interestingly, the compliance of the design in Fig. 2.14f is 3% lower! This counter-intuitive result is not entirely unexpected. As White and Voronin (2019) noted, in some cases—such as when the aspect ratio of the 2D cantilever is around 2.5 (as in Fig. 2.14a)—asymmetric designs can outperform symmetric ones, exhibiting lower compliance, even when the boundary and load conditions are totally symmetric.

Another reason for obtaining distinct topologies after introducing random voids in the initial design model is that the BESO process, like other commonly used

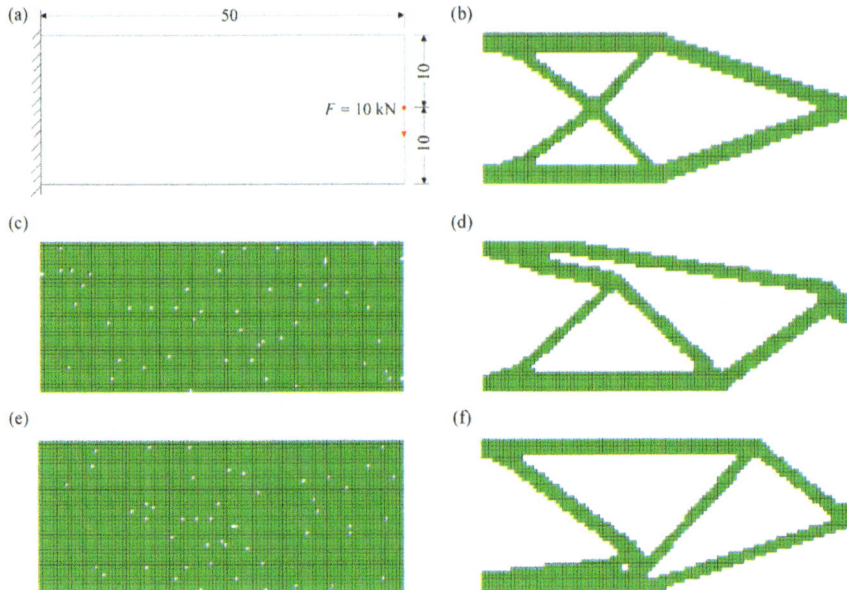

Fig. 2.14 Multiple designs for a cantilever: **a** boundary and load conditions; **b** optimized design from the conventional BESO method; **c** and **e** introducing random voids to the initial finite element model; **d** and **f** alternative designs generated after random voids are introduced to the initial finite element model (reprinted from Xie et al. 2019, with permission from authors)

topology optimization algorithms, is history dependent. In particular, the BESO method averages the element sensitivity with its historical value (Huang and Xie 2007). This feature can be considered an advantage when we want to enhance design diversity in topology optimization results.

2.6.2 Penalizing Element Sensitivities by Random Coefficients

In Sect. 2.4, sensitivities of elements within a selected region are penalized by a constant coefficient. Here, we introduce a different technique that penalizes sensitivities of all elements in the design domain using random coefficients:

$$\overline{\alpha}_i = \beta_i \tilde{\alpha}_i, \ \beta_i \in [1 - \varepsilon, 1 + \varepsilon] \tag{2.1}$$

where $\overline{\alpha}_i$, $\tilde{\alpha}_i$, and β_i are the penalized sensitivity, the original sensitivity, and the penalty coefficient of the ith element, respectively. The penalty coefficient β_i is randomly selected from a specific range $[1 - \varepsilon, 1 + \varepsilon]$, where ε is a prescribed value, such as 0.15. The original sensitivity $\tilde{\alpha}_i$ is calculated using Eq. (4.14) for the BESO method. If a different topology optimization method is used, an equivalent element sensitivity can be used for $\tilde{\alpha}_i$.

Note that the penalization of element sensitivities is conducted in each iteration. As a result, the ranking of elements based on their sensitivities is altered, which affects the removal or addition of 'marginal' elements that originally have relatively low sensitivities. However, elements with relatively high sensitivities still have a higher chance of being retained in the final design and thus the obtained solution is most likely to be structurally efficient.

Figure 2.15a shows a deep beam with a pin support at one end and a roller at the other. Using the BESO method, we obtain a conventional design in Fig. 2.15b. By setting ε to 0.15, we penalize the sensitivity of each element by a coefficient randomly selected from the range [0.85, 1.15]. Due to the random nature of the penalization, a new solution can be obtained every time the full topology optimization process is run. Two such solutions are presented in Fig. 2.15c, d, which exhibit distinctly different geometries. Interestingly, the compliance values of both new designs are 1% *lower* than that of the original solution in Fig. 2.15b, despite the fact that the designs in Fig. 2.15c, d are significantly asymmetric and slightly asymmetric, respectively.

The technique described above can be easily implemented for 3D problems. Figure 2.16a shows a 3D long cantilever fixed on the left-hand side, with a vertical load applied at the centre of the opposite side. By setting the target volume fraction to 15% and the filter radius to 3, we obtain a symmetric design (see Fig. 2.16b) from the conventional BESO method. If the sensitivity of each element is penalized by a coefficient randomly selected from the range [0.85, 1.15], a new design can be obtained every time the full topology optimization process is run. Two such designs are shown in Fig. 2.16c, d, which exhibit distinctly different, asymmetric geometries. However,

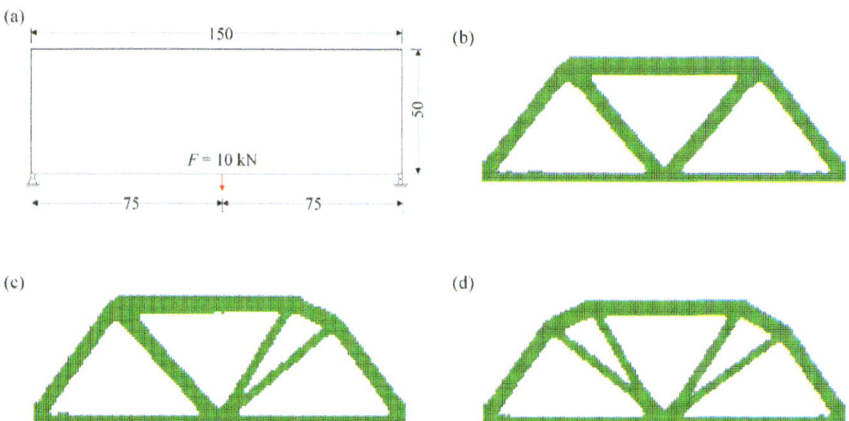

Fig. 2.15 Multiple designs for a deep beam: **a** boundary and load conditions; **b** optimized design from the conventional BESO method; **c** and **d** alternative designs generated after penalizing element sensitivities by random coefficients (reprinted from Xie et al. 2019, with permission from authors)

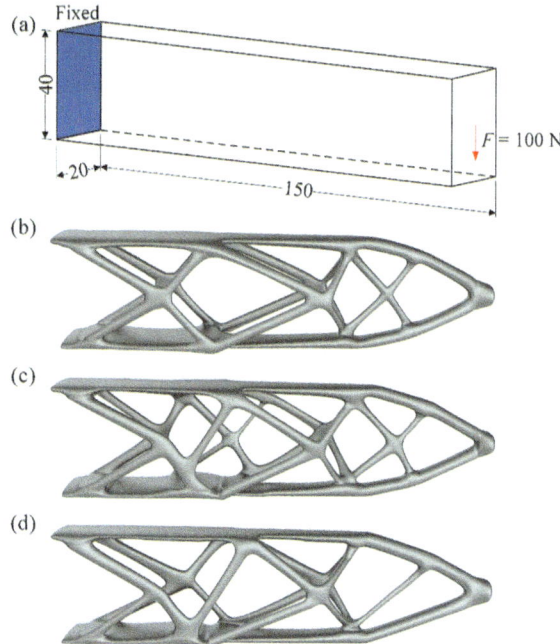

Fig. 2.16 Multiple designs for a 3D long cantilever with a target volume fraction of 15%: **a** load and support conditions; **b** optimized design from the conventional BESO method; **c** and **d** alternative designs generated by penalizing element sensitivities using random coefficients (reprinted from He et al. 2020, with permission from Elsevier)

their compliance values only increase by 0.5% and 0.4%, respectively, compared to the original design in Fig. 2.16b.

It is worth noting that, other than penalizing the element sensitivity directly, we can also generate diverse and competitive designs by penalizing the material properties of all or selected elements, using either random coefficients or a constant factor. One of the techniques discussed in the following section exploits this concept.

2.6.3 Perturbing Load and Support Conditions

We have developed another technique to introduce noise into the optimization process by randomly perturbing load and support conditions, which leads to a variety of local optima that could be considered diverse and competitive designs.

The load perturbation approach is illustrated using a short cantilever, as shown in Fig. 2.17. In this example, the actual direction of the load at the midpoint of the free end is vertically downwards. During the optimization process, we deliberately change the load direction in early iterations by a perturbation angle, η, which is randomly selected within a perturbation range $[-\varphi^L, \varphi^L]$. Starting from an initial value, such as 30°, the perturbation amplitude φ^L is progressively reduced towards zero as the optimization advances and then maintained at zero until the optimization process converges.

2.6 Introducing Randomness into the Optimization Process

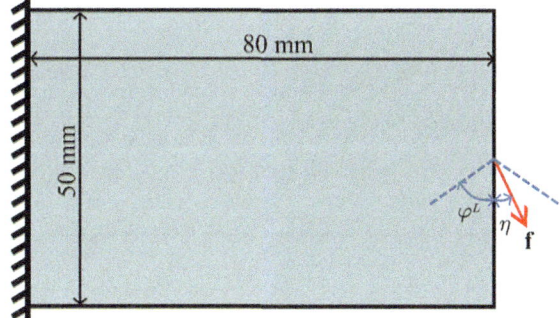

Fig. 2.17 Illustration of the load perturbation approach using a short cantilever example, where η and φ^L represent the perturbation angle and perturbation range, respectively (reprinted from Xiong et al. 2024, licensed under CC-BY 4.0)

Figure 2.18a shows the conventional BESO process where the load direction is kept constant. As expected, a symmetric design is obtained. Next, we temporarily change the load from its actual direction by a perturbation angle, η. In this case, the perturbation amplitude φ^L is initially set to 30° and the target volume fraction is 50% (see Fig. 2.18b). Due to the random nature of the load perturbation, each time the optimization process is run, a new design with different geometric features may emerge. Extensive numerical tests on the short cantilever reveal that, despite the rich diversity in the obtained forms, the compliance values of all the solutions only vary by less than 0.4% compared to that of the original BESO design (Xiong et al. 2024).

Given the history-dependent nature of the BESO and other topology optimization methods, perturbing the loads in early iterations can lead to new solutions that are geometrically diverse. Further, as the amplitude of load perturbation progressively diminishes towards zero in later iterations (see Fig. 2.18b), we ensure that the optimized designs are structurally efficient for the actual load direction.

Next, the load perturbation approach is applied to the design of a spherical shell dome shown in Fig. 2.19. The 5 m high dome has a base radius of 32 m and an oculus of 4 m radius at the top. Note that this dome is much shallower than the one shown in Fig. 2.2 which has a height of 19 m. The bottom edge of the dome is supported by pins in 16 equally spaced areas (see Fig. 2.19a). The optimized design from the conventional BESO method for the target volume fraction of 60% is given in Fig. 2.19b.

Figure 2.20 presents a variety of optimized designs after the direction of the gravity load is perturbed during early iterations of the optimization process. Specifically, when the volume fraction VF > 90%, the range of the perturbation angle is set to $[-10°, 10°]$ (i.e. $\varphi^L = 10°$); when 80% ≤ VF ≤ 90%, $\varphi^L = 8°$; when 70% ≤ VF < 80%, $\varphi^L = 5°$; when VF < 70%, φ^L is set to zero for the remainder of the optimization process until convergence is reached. Note that despite the load perturbation during early iterations, the final converged solution is for the gravity load in its actual direction. Compared to the original BESO design (see Fig. 2.19b), the change in the structural compliance of the six geometrically different designs

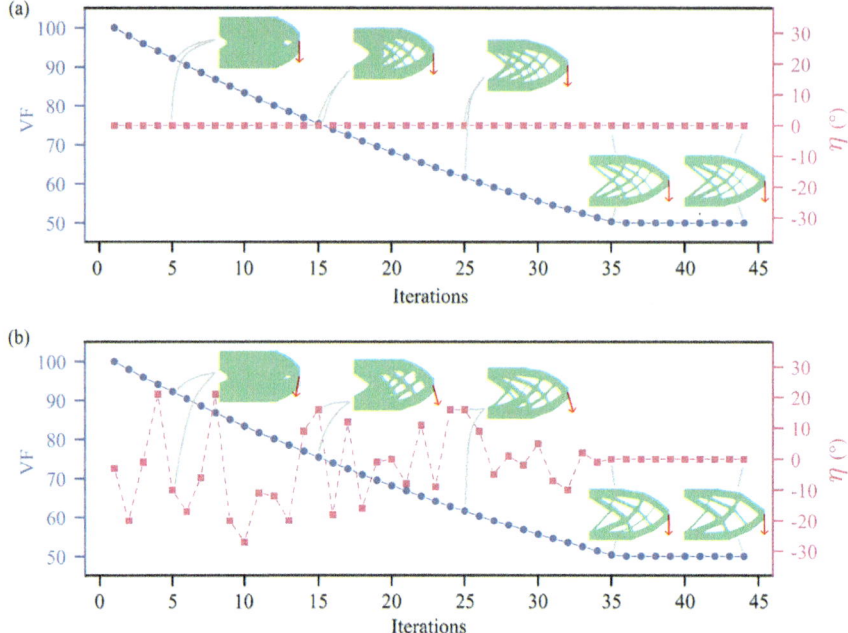

Fig. 2.18 Evolution histories of the volume fraction and perturbation angle for the short cantilever example: **a** conventional BESO process where the load direction is kept constant; **b** new optimization process with the load direction perturbed randomly within a diminishing range (reprinted from Xiong et al. 2024, licensed under CC-BY 4.0)

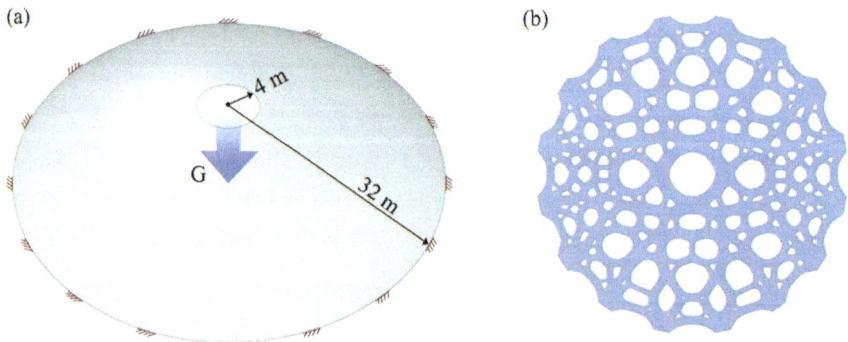

Fig. 2.19 Design of a spherical dome: **a** boundary and load conditions, where G represents gravity; **b** top view of the optimized design from the conventional BESO method (reprinted from Xiong et al. 2024, licensed under CC-BY 4.0)

2.6 Introducing Randomness into the Optimization Process

in Fig. 2.20 ranges from − 0.8% to 0.4%, which is remarkably small. These new solutions are indeed diverse and competitive designs.

It is worth noting that the optimization results shown in Figs. 2.19 and 2.20 are obtained by assuming structural symmetry about two orthogonal planes and using only a quarter of the dome in the optimization model. While this approach saves computational time, adopting a full model could further enhance both design diversity and structural performance, as demonstrated in Figs. 2.14 and 2.15 where asymmetric designs emerge.

Similar to the load perturbation approach, we may apply perturbation to the material properties, such as the Young's modulus, of elements near the supports:

$$E = (1 + \delta)E_o \qquad (2.2)$$

where E and E_o are the perturbed and original Young's moduli, respectively, and the amount of perturbation δ is randomly selected from a perturbation range $[-\varphi^S, \varphi^S]$ at each iteration. Starting from an initial value, such as 20%, φ^S is progressively reduced towards zero as the optimization advances and then is maintained at zero until the optimization process converges.

The support perturbation approach is illustrated using a column design example shown in Fig. 2.21a. The design domain measuring 60 mm in length and 120 mm in height is discretized using triangular elements with an average edge length of 0.1 mm. Three vertical forces are applied at two end points and the midpoint of the top edge, while the structure is fixed at two end points and the midpoint of the bottom edge. With the target volume fraction and the filter radius set to 30% and

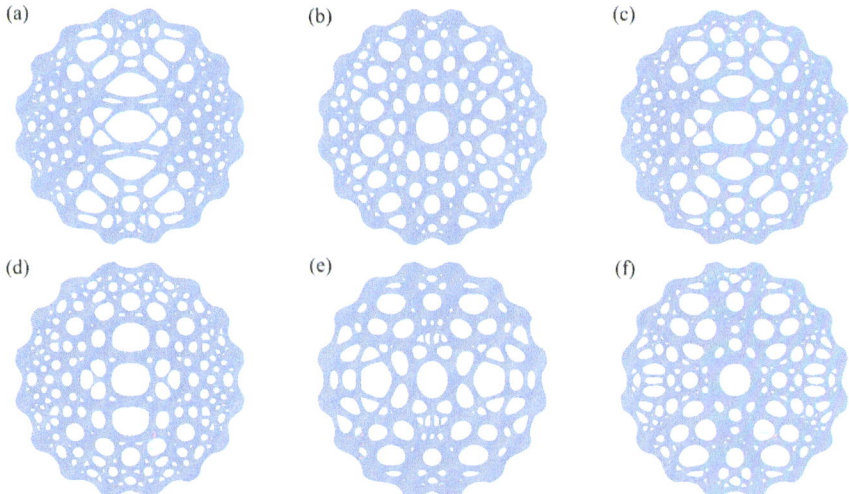

Fig. 2.20 Diverse designs for the dome obtained from the load perturbation approach (reprinted from Xiong et al. 2024, licensed under CC-BY 4.0)

1 mm, respectively, we obtain the optimized design from the conventional BESO method, as shown in Fig. 2.21b.

Next, we apply perturbations to the Young's modulus of elements in selected regions near the three supports. When the volume fraction VF $\geq$ 50%, the perturbation range is set to $[-20\%, 20\%]$ (i.e. $\varphi^S = 20\%$); when $30\% \leq$ VF $< 50\%$, $\varphi^S = 10\%$; when VF $< 30\%$, φ^S is set to zero for the remainder of the optimization process until convergence is reached.

Figure 2.22 shows a variety of designs obtained from the support perturbation approach with the perturbation region at each support being set to 2 mm × 2 mm, 4 mm × 4 mm, and 20 mm × 20 mm, respectively. While the original design in Fig. 2.21b features a vertical structural member in the central upper region, several of the new designs in Fig. 2.22 exhibit inverted V-shaped components. Compared to the original design in Fig. 2.21b, the changes in the structural compliance (ΔC) of all the new designs in Fig. 2.22 are very small, ranging from -0.359% to 0.562%.

Note that the effectiveness of the support perturbation may be influenced by the ratio of the edge length of the perturbed region to the filter radius. When the edge length of this region is smaller than the filter radius, the intended sensitivity 'noise' from the perturbation tends to be averaged out, resulting in a reduction in the diversity of solutions. Therefore, it is recommended that the edge length of the perturbed region be no less than the filter radius. In the column design example in Fig. 2.22, the edge lengths of the perturbed region have been set to 2, 4, and 20 times the filter radius, respectively.

To further enhance design diversity, we can apply perturbations to both load and support. Here, we consider a 3D pavilion design with a free-form top surface (see Fig. 2.23a). The design space measures 18 m in length, 9 m in width, and 7.6 m

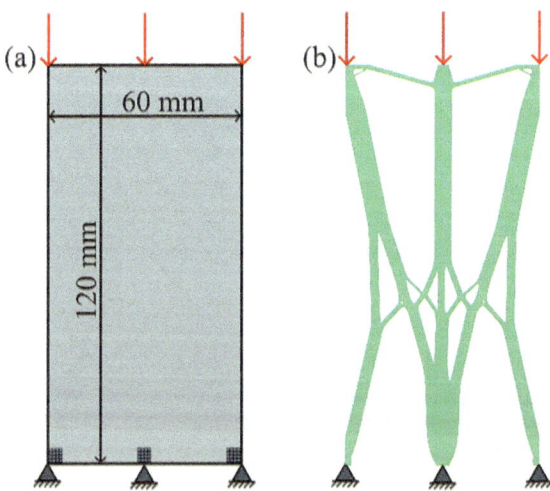

Fig. 2.21 Illustration of the support perturbation approach using a column design example: **a** support and load conditions, and selected regions near supports for applying the perturbation; **b** optimized design from the conventional BESO method (reprinted from Xiong et al. 2024, licensed under CC-BY 4.0)

2.6 Introducing Randomness into the Optimization Process

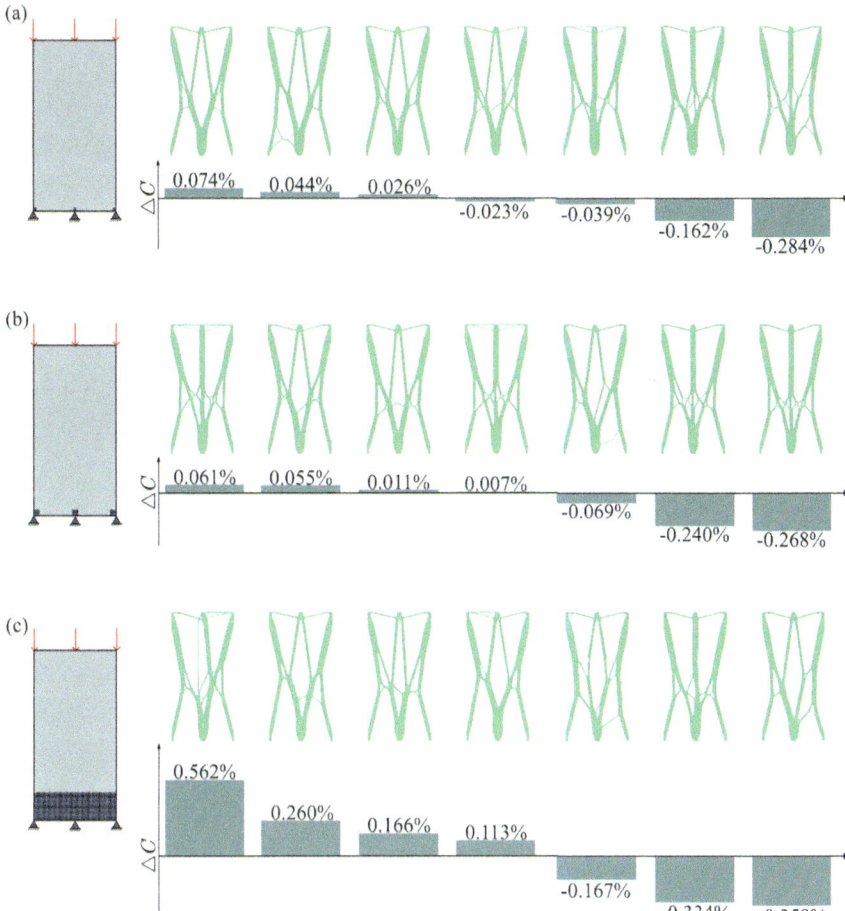

Fig. 2.22 Diverse designs for the column obtained from the support perturbation approach with a perturbation region of different sizes at each support: **a** 2 mm × 2 mm; **b** 4 mm × 4 mm; **c** 20 mm × 20 mm (reprinted from Xiong et al. 2024, licensed under CC-BY 4.0)

in height. The curved top layer, with a thickness of 0.3 m, is specified as non-design domain. The bottom layer of 0.9 m thickness is set as the region for support perturbation. A uniformly distributed vertical pressure load is applied to the top surface, while the bottom surface of the structure is fixed. With the target volume fraction and filter radius set to 8% and 0.9 m, respectively, the optimized design obtained from the conventional BESO method is shown in Fig. 2.23b, featuring 10 columns distributed in a doubly symmetric pattern.

Figure 2.24 shows a variety of new designs created after applying random perturbations within specified ranges to the load in the xz plane (see Fig. 2.23a for the coordinate system) and to the support simultaneously during early iterations of the

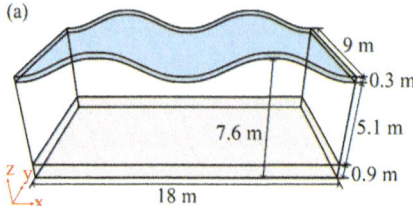 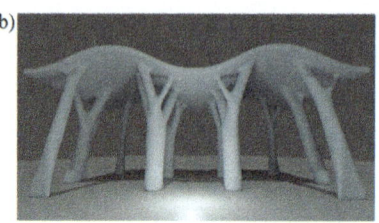

Fig. 2.23 Design of a pavilion: **a** non-design domain (curved top surface with a thickness of 0.3 m) and support perturbation region (bottom layer with a thickness of 0.9 m); **b** optimized design from the convention BESO method (reprinted from Xiong et al. 2024, licensed under CC-BY 4.0)

topology optimization process. The images on the right-hand side of Fig. 2.24 provide top views of the individual columns. The main differences among these designs are the number, positions, and shapes of columns supporting the curved roof. Since the load perturbation is applied in the xz plane to the originally vertical force, the perturbed load has a force component in the x direction, leading to multiple columns branching out in that direction, including columns 1, 2, 3, 4, 5, and 8 in Fig. 2.24a; columns 4, 7, 9, and 10 in Fig. 2.24b; columns 2, 3, 4, 7, and 8 in Fig. 2.24c; and columns 1, 3, 4, 5, 6, 7, and 8 in Fig. 2.24d.

Compared to the original design in Fig. 2.23b, the changes in the structural compliance (ΔC) of the four new designs in Fig. 2.24a–d are -2.27%, -1.81%, -1.14%, and -0.10%, respectively. This indicates that the diverse designs in this case have slightly improved stiffness even though they are asymmetric.

2.7 Explicitly Controlling the Structural Complexity

In all the techniques discussed so far in this chapter, although the designer can exert some influence over the optimization results, the design outcomes are largely unpredictable and sometimes random (as shown in Sect. 2.6). However, for some architectural and engineering applications, the designer may need precise control over various geometric features, such as the number and size of rooms in a building. In this section, we introduce methods for explicitly controlling the numbers and sizes of cavities and tunnels in topology optimization.

Figure 2.25 illustrates the basic concepts of cavities and tunnels. A cavity is a fully enclosed void within a structure (see Fig. 2.25a), whereas a tunnel is a pathway running through a structure (see Fig. 2.25b). Enclosed voids are commonly observed in topology-optimized 2D structures (see Fig. 2.25c) and such voids are also regarded as cavities. However, enclosed voids are seldomly encountered in 3D optimized results, as most of the cavities seen in 2D models become tunnels in their 3D counterparts (see Fig. 2.25d). Due to the length-scale constraint introduced by, for example, the filter radius, topology-optimized 3D structures are usually truss-like, open-walled designs with multiple tunnels (Sigmund et al. 2016).

2.7 Explicitly Controlling the Structural Complexity

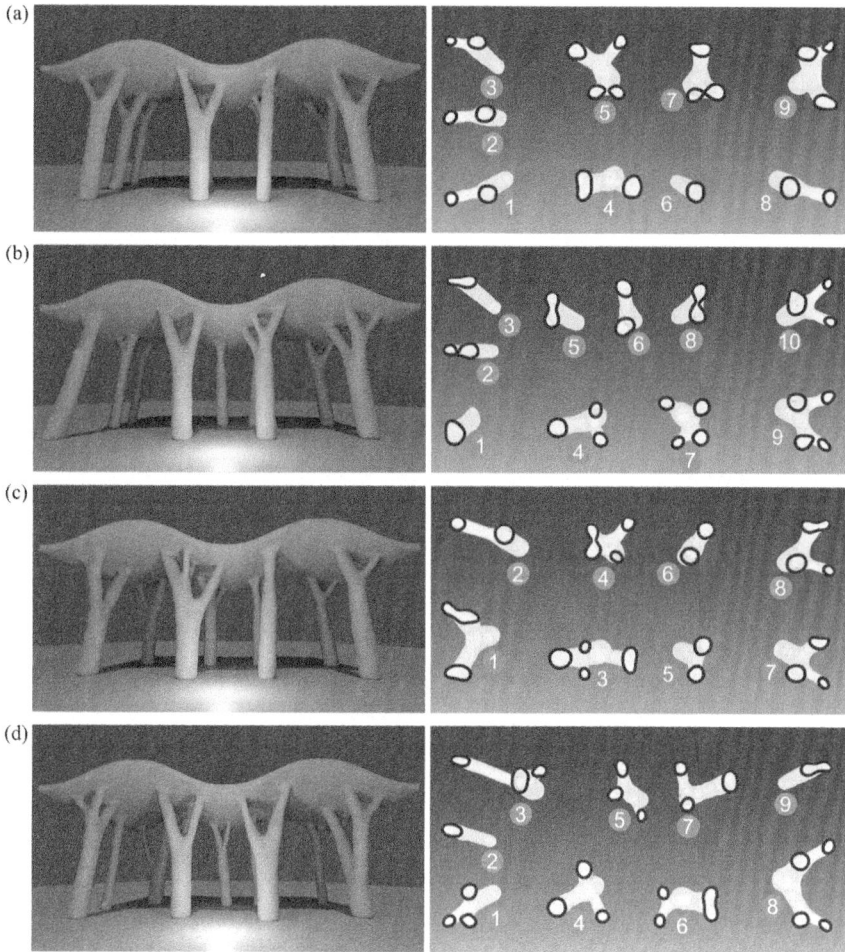

Fig. 2.24 Diverse designs for the pavilion obtained from applying load perturbation in the xz plane and support perturbation simultaneously (reprinted from Xiong et al. 2024, licensed under CC-BY 4.0)

It is worth noting that for some engineering applications, cavities should be avoided if the structure is to be fabricated using certain additive manufacturing processes such as selective laser melting and fused deposition modelling (Ngo et al. 2018), as the unmelted powder or the support material needs to be removed from the printed part. To this end, we have developed a topology optimization technique to eliminate such cavities while maintaining the structural performance (Xiong et al. 2020).

Although there has been considerable research on explicitly controlling topology in structural optimization (e.g., Zhang et al. 2017a,b; Zhao et al. 2020; Han et al. 2021, 2022; Liang et al. 2022; Wang et al. 2022), the previous methods are unable to

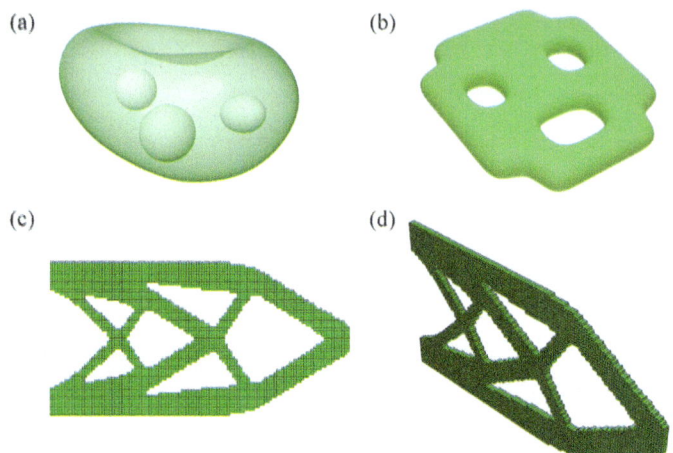

Fig. 2.25 Illustration of cavities and tunnels: **a** a 3D structure with three cavities; **b** a 3D structure with three tunnels; **c** a 2D structure with six cavities; **d** the 3D counterpart with six tunnels (**a** reprinted from Xiong et al. 2020, with permission from Elsevier; **c** and **d** reprinted from He et al. 2023, with permission from authors)

Fig. 2.26 Optimized designs of a 3D cantilever with prescribed target numbers of tunnels ($\tilde{T}$) and cavities ($\tilde{C}$): **a** $\tilde{T} = 0$ and $\tilde{C} = 0$; **b** $\tilde{T} = 5$ and $\tilde{C} = 0$; **c** $\tilde{T} = 10$ and $\tilde{C} = 0$ (reprinted from He et al. 2023, with permission from authors)

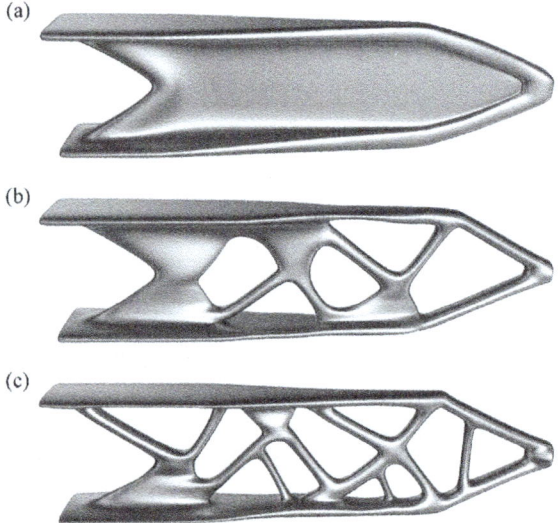

precisely control the number and size of tunnels that appear constantly in topology optimized 3D models. To address this challenge, we have recently developed a 'hole-filling' method that can explicitly control the numbers and sizes of tunnels and cavities (He et al. 2023). In the hole-filling method, excess cavities are filled with solid material and excess tunnels are blocked by building plate-like patches. Once the target numbers of tunnels and cavities are achieved, a topology-preserving optimization

2.7 Explicitly Controlling the Structural Complexity

technique based on a thinning algorithm (He et al. 2022) is used to find the final design.

We apply the hole-filling method to the 3D cantilever given in Fig. 2.16a. During the optimization process, the numbers of tunnels and cavities are precisely controlled. The minimum volumes of each cavity and tunnel are set to 2% and 0.2% of the cuboid design domain, respectively. Figure 2.26 shows the results for different target numbers of tunnels ($\tilde{T}$) and cavities ($\tilde{C}$). When $\tilde{T} = 0$ and $\tilde{C} = 0$, we obtain a plate-like, closed-walled structure (see Fig. 2.26a) whose compliance is 7% *lower* than that of the original truss-like design (see Fig. 2.16b). As noted by Sigmund et al. (2016), plate-like or closed-walled structures are often stiffer than truss-like designs. When $\tilde{T} = 5$ and $\tilde{T} = 10$, the optimized designs shown in Fig. 2.26b, c contain smaller regions of closed walls, but the compliance values of both structures are still 3% lower than that of the original truss-like design in Fig. 2.16b.

Further, we can apply different topological constraints to selected regions of the design domain. Figure 2.27a shows the optimized design of the 3D cantilever (see Fig. 2.16a) without topological constraints when the target volume fraction is set to 35%. Then, we divide the cuboid design domain into two regions (see Fig. 2.27b). Figure 2.27c presents the optimized design when $\tilde{C} = 3$ and $\tilde{T} = 3$ are specified in the grey and red regions, respectively. Figure 2.27d shows the optimized design when $\tilde{C} = 2$ and $\tilde{T} = 4$ are specified in the grey and red regions, respectively. Compared to the original solution (see Fig. 2.27a), the compliance values of the

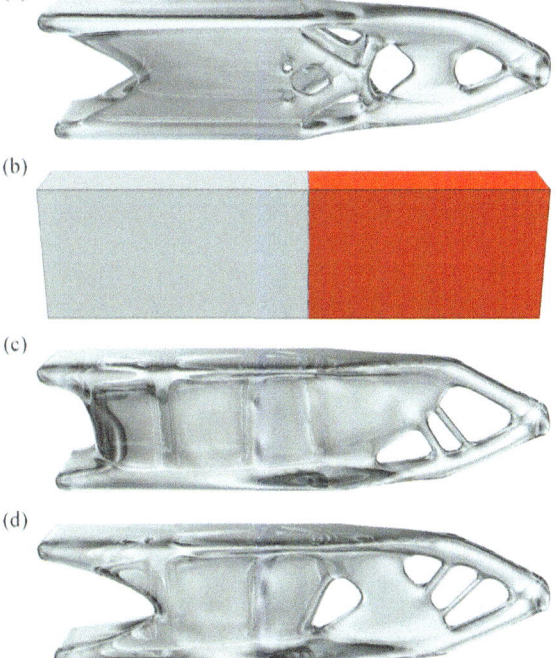

Fig. 2.27 Controlling both cavities and tunnels in a 3D cantilever with target volume fraction of 35%: **a** conventional optimized design without topological constraints; **b** two selected regions of the design domain; **c** optimized design with $\tilde{C} = 3$ in the grey region and $\tilde{T} = 3$ in the red region; **d** optimized design with $\tilde{C} = 2$ in the grey region and $\tilde{T} = 4$ in the red region (reprinted from He et al. 2023, with permission from authors)

two new designs in Fig. 2.27c, d have increased by 5% and 8%, respectively. These slight reductions in structural performance have enabled us to achieve two distinctly different solutions that precisely meet the requirements of the prescribed numbers of cavities and tunnels in selected regions, while also satisfying the aforementioned constraints on the minimum volumes of each cavity and tunnel.

2.8 Conclusion

In this chapter, we have presented a variety of techniques that can be easily implemented to achieve geometrically diverse and structurally competitive designs. Although we have developed these techniques and examples based on the BESO method and for continuum structures, the same concepts can be applied to other topology optimization methods and for discrete structures (see Cai et al. 2021). We have demonstrated that these techniques can be used to readily obtain multiple solutions with distinctly different geometries but similar compliance values. It is envisaged that these techniques would be applicable to some other design objectives or constraints such as the natural frequency (a global property of a structure). However, for optimization problems with constraints that are highly sensitive to variations of local geometries, such as stress and buckling, achieving diverse and competitive designs could be more challenging. Further research in this area would be interesting and worthwhile.

References

Cai, Q., He, L., Xie, Y., Feng, R., and Ma, J. (2021) Simple and effective strategies to generate diverse designs for truss structures. *Struct.* **32**, 268–278.
Cui, C., Ohmori, H. and Sasaki, M. (2003) Computational morphogenesis of 3D structures by extended ESO method. *J. Int. Assoc. Shell Spat. Struct.* **44**, 51–61.
Han, H., Guo, Y., Chen, S. and Liu, Z. (2021) Topological constraints in 2D structural topology optimization. *Struct. Multidisc. Optim.* **63**, 39–58.
Han, H., Wang, C., Zuo, T. and Liu, Z. (2022) Inequality constraint on the maximum genus for 3D structural compliance topology optimization. *Sci. Rep.* **12**, 16185.
He, Y., Cai, K., Zhao, Z.-L. and Xie, Y. M. (2020) Stochastic approaches to generating diverse and competitive structural designs in topology optimization. *Finite Elem. Anal. Des.* **173**, 103399.
He, Y., Zhao, Z.-L., Cai, K., Kirby, J., Xiong, Y. and Xie, Y. M. (2022) A thinning algorithm based approach to controlling the structural complexity in topology optimization. *Finite Elem. Anal. Des.* **207**, 103779.
He, Y., Zhao, Z.-L., Lin, X. and Xie, Y. M. (2023) A hole-filling based approach to controlling structural complexity in topology optimization. *Comput. Methods Appl. Mech. Eng.* **416**, 116391.
Huang, X. and Xie, Y. M. (2007) Convergent and mesh-independent solutions for the bi-directional evolutionary structural optimization method. *Finite Elem. Anal. Des.* **43**, 1039–1049.
Huang, X. and Xie, Y. M. (2010) *Evolutionary Topology Optimization of Continuum Structures: Methods and Applications.* Chichester: John Wiley & Sons.

References

Lazarov B. S. and Sigmund O. (2011) Filters in topology optimization based on Helmholtz-type differential equations. *Int. J. Numer. Methods Eng.* **86**, 765–81.

Liang, Y., Yan, X. and Cheng, G. (2022) Explicit control of 2D and 3D structural complexity by discrete variable topology optimization method. *Comput. Methods Appl. Mech. Eng.* **389**, 114302,

Nervi, P. L. (2018) *Aesthetics and Technology in Building: The Twenty-First-Century Edition*. Champaign: University of Illinois Press.

Ngo, T. D., Kashani, A., Imbalzano, G., Nguyen, K. T. Q. and Hui, D. (2018) Additive manufacturing (3D printing): A review of materials, methods, applications and challenges. *Compos. B Eng.* **143**, 172–196.

Rieser, J. and Zimmermann, M. (2023) Towards closed-walled designs in topology optimization using selective penalization. *Struct. Multidisc. Optim.* **66**, 158.

Saha, P. K. (2000) *Aluminum Extrusion Technology*. Materials Park: ASM International.

Sasaki, M. (2005) *Flux Structure*. Tokyo: TOTO.

Sigmund, O., Aage, N. and Andreassen, E. (2016) On the (non-)optimality of Michell structures. *Struct. Multidisc. Optim.* **54**, 361–373.

Sigmund O. and Petersson, J. (1998) Numerical instabilities in topology optimization: A survey on procedures dealing with checkerboards, mesh-dependencies and local minima. *Struct. Optim.* **16**, 68–75.

Volk, M., Yuksel O., Baran, I., Hattel, J. H., Spangenberg, J. and Sandberg, M. (2022) Cost-efficient, automated, and sustainable composite profile manufacture: A review of the state of the art, innovations, and future of pultrusion technologies. *Compos. B Eng.* **246**, 110135.

Wang, B., Zhou, Y., Tian, K. and Wang, G. (2020) Novel implementation of extrusion constraint in topology optimization by Helmholtz-type anisotropic filter. *Struct. Multidisc. Optim.* **62**, 2091–2100.

Wang, Q., Han, H., Wang, C. and Liu, Z. (2022) Topological control for 2D minimum compliance topology optimization using SIMP method. *Struct. Multidisc. Optim.* **65**, 38.

White, D. A. and Voronin, A. (2019) A computational study of symmetry and well-posedness of structural topology optimization. *Struct. Multidisc. Optim.* **59**, 759–766.

Xie, Y. M. (2022) Generalized topology optimization for architectural design. *Archit. Intell.* **1**, 2.

Xie, Y. M., Yang, K., He, Y., Zhao, Z.-L. and Cai, K. (2019). How to obtain diverse and efficient structural designs through topology optimization. *Proc. IASS Annu. Symp.*, Barcelona, 7–10 October 2019.

Xiong, Y., Yao, S., Zhao, Z.-L. and Xie, Y. M. (2020) A new approach to eliminating enclosed voids in topology optimization for additive manufacturing. *Addit. Manuf.* **32**, 101006.

Xiong, Y., Lu, H. and Xie, Y. M. (2024) Perturbation approaches to achieving diverse and competitive designs in topology optimisation. *Struct.* **68**, 107183.

Xiong, Y., Lu, H., Yan, X., He, Y. and Xie, Y. M. (2025) Achieving diverse morphologies using three-field BESO with variable-radius filter. *Eng. Struct.* **322**, 119049.

Yan, X., Bao, D., Zhou, Y., Xie, Y. M. and Cui, T. (2022) Detail control strategies for topology optimization in architectural design and development. *Front. Archit. Res.* **11**, 340–356.

Yan, X., Xiong, Y., Bao, D. W., Xie, Y. M. and Peng, X. (2023) A multi-volume constraint approach to diverse form designs from topology optimization. *Eng. Struct.* **279**, 115525.

Yang, K., Zhao, Z.-L., He, Y., Zhou, S., Zhou, Q., Huang, W. and Xie, Y. M. (2019) Simple and effective strategies for achieving diverse and competitive structural designs. *Extreme Mech. Lett.* **30**, 100481.

Zhang, W., Liu, Y., Wei, P., Zhu, Y. and Guo, X. (2017a) Explicit control of structural complexity in topology optimization. *Comput. Methods Appl. Mech. Eng.* **324**, 149–169.

Zhang, W., Zhou, J., Zhu, Y. and Guo, X. (2017b) Structural complexity control in topology optimization via moving morphable component (MMC) approach. *Struct. Multidisc. Optim.* **56**, 535–552.

Zhao, Z.-L., Zhou, S., Cai, K. and Xie, Y. M. (2020) A direct approach to controlling the topology in structural optimization. *Comput. Struct.* **227**, 106141.

Open Access This chapter is licensed under the terms of the Creative Commons Attribution 4.0 International License (http://creativecommons.org/licenses/by/4.0/), which permits use, sharing, adaptation, distribution and reproduction in any medium or format, as long as you give appropriate credit to the original author(s) and the source, provide a link to the Creative Commons license and indicate if changes were made.

The images or other third party material in this chapter are included in the chapter's Creative Commons license, unless indicated otherwise in a credit line to the material. If material is not included in the chapter's Creative Commons license and your intended use is not permitted by statutory regulation or exceeds the permitted use, you will need to obtain permission directly from the copyright holder.

Chapter 3
Redefining the Design Domain

Structural topology optimization is usually conducted within a predefined, fixed design domain. In this chapter, we demonstrate that the design domain can be flexibly redefined, serving as an effective *driver* for design innovation. By strategically exploring alternative design domains, we can generate a diverse array of geometrically distinct and structurally efficient designs, or create unusual behaviour such as mechanical non-reciprocity.

3.1 Introduction

Traditionally, topology optimization is performed within a prescribed, fixed design domain, and selecting this domain may seem like an obvious, trivial task. However, through a variety of examples, this chapter shows that by judiciously redefining the design domain, we can create a wide range of innovative and efficient structures. Various strategies are presented in this chapter, including exploring alternative design domains, fixing part of the design space, designating part of the design space as a prohibited region, embedding a preferred pattern in the design domain, selecting an adequate design domain, and using an adaptive design domain. Additionally, we show that novel non-reciprocal structures can be achieved by introducing gaps within the design domain.

3.2 Exploring Alternative Design Domains

Changing the design domain can have a significant impact on the resulting structural form. This simple approach provides a great opportunity to the engineer when collaborating with an architect or a client, particularly at the early stage of a project when details of the design are undecided yet. By exploring various design domains, the engineer can create multiple structurally efficient and geometrically different solutions. One of these designs may align with the client's preferences and meet functional or other requirements.

Figure 3.1 shows an example of the first set of designs we proposed to our collaborator, a leading bridge design firm, for a two-span footbridge made of steel and concrete. At the conceptual design stage, we assumed a uniformly distributed load on the deck, with the bottom surface of the three piers fixed to the ground. By simply varying the design domain while keeping the load and support conditions unchanged (see Fig. 3.1a), we quickly generated a range of distinctly different preliminary designs (see Fig. 3.1b) for further discussion with our collaborator. Each of the four designs meets the basic requirements of a footbridge, but the final choice may depend on various factors, such as aesthetics, construction cost, and the client's preferences. For this project, we employed the multi-material bi-directional evolutionary structural optimization (BESO) method developed by Li and Xie (2021a, b), which judiciously put steel in tensile regions and concrete in compressive areas to achieve structural efficiency.

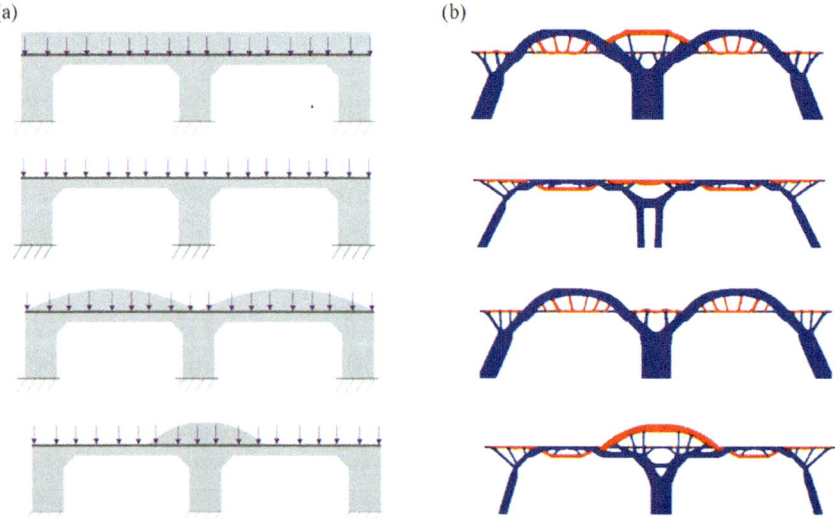

Fig. 3.1 Exploring alternative design domains for a footbridge: **a** different design domains (grey areas) for the same load and support conditions; **b** optimized distributions of concrete (blue areas) and steel (red areas) obtained by applying the multi-material topology optimization method (reprinted from Xie 2022, licensed under CC-BY 4.0)

3.3 Fixing Part of the Design Space

In architectural and engineering designs, it is often necessary to retain certain parts of a structure—such as the deck of a bridge—even if they are not essential for structural performance. Additionally, a designer may wish to incorporate specific geometric features within the structure. In such cases, parts of the structure can be designated as *non-design domain*, also known as 'passive solid domain', so that these areas remain unchanged during the topology optimization process.

Consider a cantilever beam with a vertical point load applied at the bottom right corner (see Fig. 3.2a). Without specifying a non-design domain, we obtain the solution shown in Fig. 3.2b. However, if we designate a preferred pattern—such as the A-shaped, B-shaped, or C-shaped area—as the non-design domain, we can generate a new topology of the same total volume that preserves the specified pattern in the final design (see Fig. 3.2c). Although the new designs are geometrically distinct from the original solution (see Fig. 3.2b), they remain highly competitive in terms of structural performance. Specifically, compared to the original design, the compliance values of the new designs in Fig. 3.2c increase by only 3% (A-shaped), 7% (B-shaped), and 2% (C-shaped), respectively (Yang et al. 2019).

Next, we show an interesting example by Li et al. (2022), which was inspired by an architectural design test case of Ameba—a topology optimization tool based on the BESO method (Zhou et al. 2018; XIE Technologies 2024). This structure has a curved top surface subjected to a uniformly distributed load (see Fig. 3.3a). Within a large block of design space, an A-shaped part is specified as non-design domain. The bottom surface of the non-design domain is fixed to the ground (see Fig. 3.3b). While inefficient material surrounding the non-design domain can be removed freely, the A-shaped part is kept unchanged throughout the optimization process, resulting in the visually appealing final design shown in Fig. 3.3c.

In the examples shown in Figs. 3.2 and 3.3, the non-design domain is prescribed before the optimization process begins and then remains fixed throughout the iterative

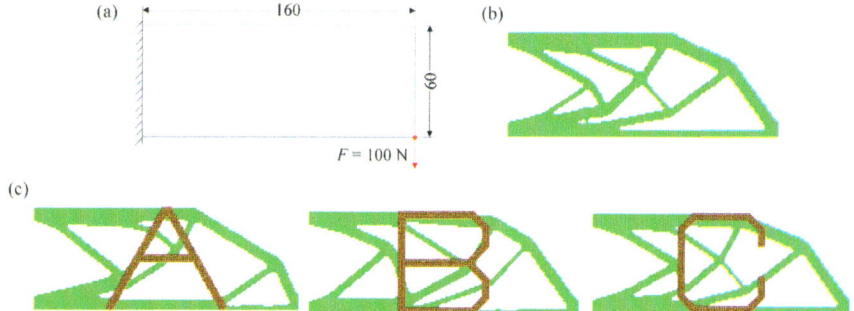

Fig. 3.2 A cantilever beam with a prescribed pattern: **a** load and support conditions; **b** optimized design without non-design domain; **c** optimized designs with prescribed, fixed patterns (reprinted from Yang et al. 2019, with permission from Elsevier)

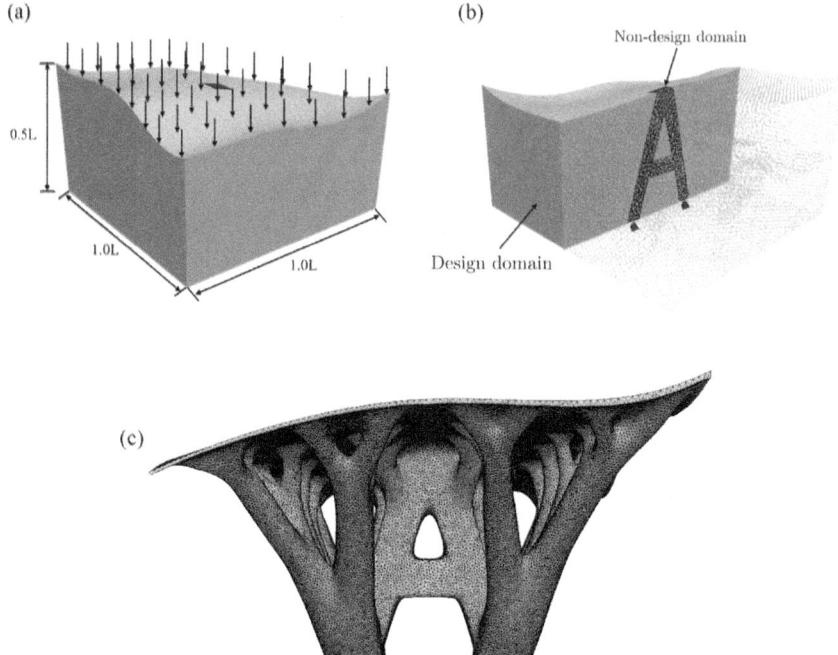

Fig. 3.3 A free-form structure with an A-shaped non-design domain: **a** load condition and design space; **b** support condition and non-design domain; **c** optimized design (reprinted from Li et al. 2022, with permission from Elsevier)

process. However, it is sometimes desirable to increase design flexibility by allowing these prescribed parts or patterns to change their locations and orientations in the design space during the optimization process. This approach may result in more efficient solutions. For more details, interested readers are referred to Zhang and Zhou (2020), which provides an in-depth discussion of the 'feature-driven' method for structural optimization.

Other commonly used strategies for directly controlling geometrical patterns in topology optimization include imposing symmetric or periodic constraints on the structure (Huang and Xie 2008; He and Xie 2024). These strategies can be easily implemented in topology optimization tools and are highly useful for architectural and engineering applications.

3.4 Setting Part of the Design Space as a Prohibited Region

A *prohibited region*, also known as *passive void domain*, is an empty space that must remain unoccupied by material during the optimization process. Setting part of the design space as a passive void domain can significantly affect the design outcome. An example is presented in Fig. 3.4. To design a bridge using topology optimization, we set the top layer as non-design domain for the deck and the cuboid below as design domain, with the four bottom corners fixed and a uniformly distributed load applied to the top surface (see Fig. 3.4a). The optimized result is given in Fig. 3.4b. However, this design does not meet the basic functional requirement of a bridge, as the two bars at the bottom would obstruct traffic flow below the deck. To address this issue, we create a narrow gap below the deck and set this space as a passive void domain (see Fig. 3.4c). This adjustment produces a conceptually different design in Fig. 3.4d, which represents an interesting and elegant arch bridge. While the compliance of the design in Fig. 3.4d is 14% higher than that of the original solution in Fig. 3.4b, the structurally less efficient design in Fig. 3.4d is more likely to be adopted as a viable bridge design.

If one wishes to further enhance the structural performance and at the same time provide adequate clearance underneath the deck, an effective approach—commonly employed in many real bridges—is to allow structural members to form *above* the deck. To achieve such a new bridge design, an expanded design domain needs to be prescribed. Alternatively, the design domain should adapt automatically according to the structural requirement. In Sect. 3.7, we use the so-called adaptive design domain method to solve such a bridge design problem, allowing structural components to 'grow' freely beyond the initial design domain and achieving structurally more efficient solutions.

In the above example, the passive void domain is prescribed before the optimization process begins and then remains fixed throughout the iterative process. It is possible to further enhance design flexibility by allowing the prescribed void to

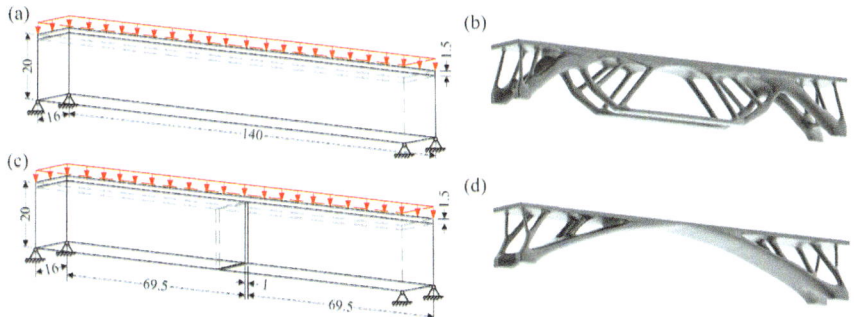

Fig. 3.4 Setting part of the design space as a passive void domain: **a** and **c** load and support conditions, and different design domains; **b** and **d** optimized designs without and with a passive void domain (reprinted from Yang et al. 2019, with permission from Elsevier)

change its size, shape, and location during the optimization process, which may result in more efficient structural designs (Clausen et al. 2014; Zhang and Zhou 2020).

3.5 Embedding a Geometric Pattern in the Design Domain

Geometric patterns have been widely used in architecture across different cultures throughout history. For example, Islamic artists and architects have adorned buildings and monuments with intricate geometric patterns for centuries (Bonner 2017).

The visual beauty of geometric patterns can be incorporated into topology optimization results by embedding such patterns in the design domain. We illustrate this concept using a spherical shell example given in Fig. 3.5 (Meng et al. 2023). The shell's radius of curvature, length, width, and thickness are 20 m, 10 m, 10 m, and 0.1 m, respectively. The shell is fixed at the four bottom corners and subjected to gravity load (see Fig. 3.5b). Using the BESO method (Huang and Xie 2010), we obtain a conventional topology optimization result shown in Fig. 3.5c when the volume fraction is set to 80%. While this design is structurally efficient, its form may not appeal to the architect aesthetically.

Figure 3.5d presents an alternative solution after an array of circular holes is embedded in the design domain. During the optimization process, only material within these holes is allowed to be removed. In other words, the area outside the prescribed geometric pattern is treated as non-design domain. Despite the distinct visual differences between the two designs in Fig. 3.5c, d, the compliance of the new design incorporating the geometric pattern has increased by only 0.2% compared to the conventional design. This indicates that embedding a geometric pattern in the design domain may have a small or negligible impact on the structural performance.

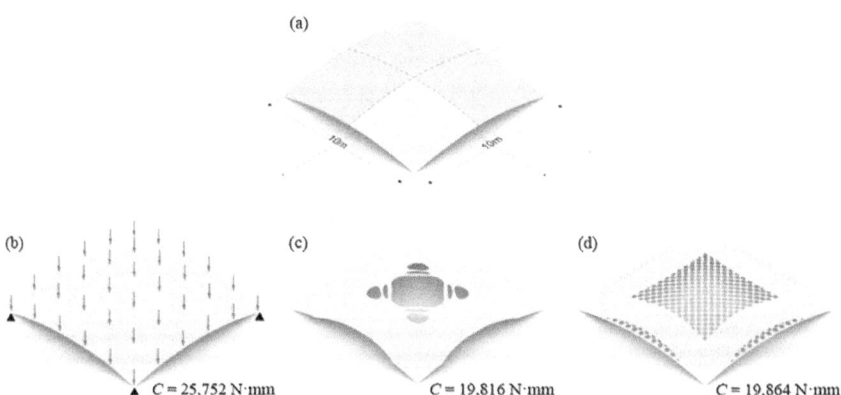

Fig. 3.5 Designing a spherical shell under gravity load: **a** shell dimensions; **b** load and support conditions; **c** result from conventional topology optimization; **d** optimized design incorporating an embedded geometric pattern (reprinted from Meng et al. 2023, with permission from Elsevier)

3.5 Embedding a Geometric Pattern in the Design Domain

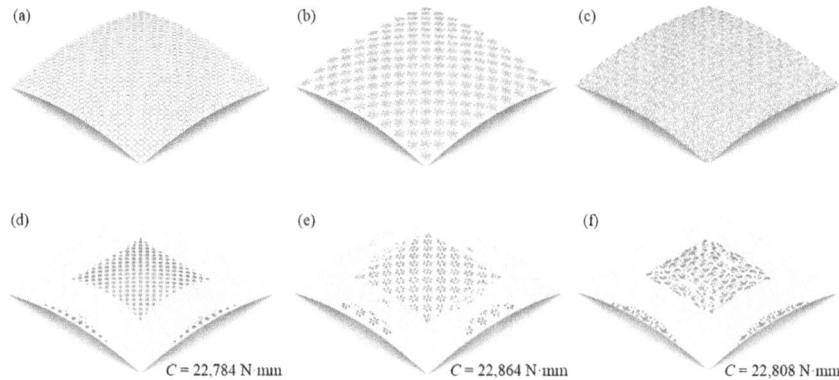

Fig. 3.6 Designing a spherical shell with various embedded geometric patterns in the design domain: **a** circular-hole pattern; **b** flower pattern; **c** T-shaped pattern; **d–f** optimized designs corresponding to the circular-hole, flower, and T-shaped patterns, respectively (reprinted from Meng et al. 2023, with permission from Elsevier)

The embedded geometric pattern tends to emerge within or near the void regions of the conventional design, leaving the solid regions largely unaffected.

Next, we examine the effect of applying different geometric patterns to the design domain. In addition to the circular-hole pattern shown in Fig. 3.6a, a flower pattern (see Fig. 3.6b) and a T-shaped pattern (see Fig. 3.6c) are considered. With the volume fraction set to 90%, we obtain three new optimized designs in Fig. 3.6d–f. Although these three designs exhibit distinctly different geometric patterns, their compliance values differ by less than 0.4%.

In the following example, we consider a free-form shell with a large, fixed opening on the top, as illustrated in Fig. 3.7a (Meng et al. 2023). This structure is inspired by the Teshima Art Museum, designed by the renowned Japanese architect Ryue Nishizawa (Nishizawa 2010).

The shell's overall dimensions are approximately 60 m × 40 m × 5 m, with a thickness of 0.25 m in the outer region and 0.15 m in the inner region. The shell is subjected to gravity load, with its bottom edge fixed. Figure 3.7b shows an array of circular holes predefined across the design domain to form an embedded pattern. The designer can freely select the locations, sizes, and shapes of these holes based on their aesthetical preferences and other considerations such as natural lighting. During the subsequent topology optimization process, only material within these holes is allowed to be removed.

Figure 3.7c shows a rendering of the optimized design with the target volume fraction (VF) set to 90%. Solutions for various target volume fractions, ranging from 98% to 90%, are displayed as insets within the figure. It is observed that most of the additional, smaller openings emerge near the supporting edge of the shell. The rendering of the optimized design shows that the distributed circular openings are aesthetically pleasing and may also be beneficial for natural airflow and daylight.

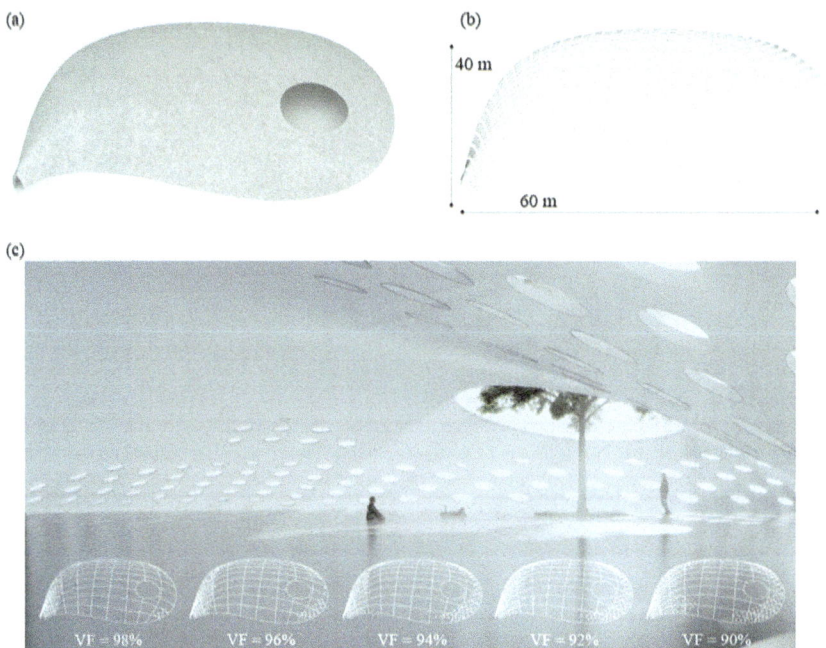

Fig. 3.7 Designing a free-form shell with a large opening and an embedded pattern of smaller holes: **a** a fixed large opening on the shell; **b** an embedded pattern of small circular holes; **c** rendering of the optimized design for VF = 90%, with insets showing optimized designs for various target volume fractions (reprinted from Meng et al. 2023, with permission from Elsevier)

3.6 Selecting an Adequate Design Domain

For some engineering applications, defining an appropriate design domain is not an obvious task, and making an improper choice can significantly hinder the structural performance of the optimization result. To demonstrate this point, we consider the design of a hinge arm for the load and boundary conditions shown in Fig. 3.8.

This seemingly straightforward design task originates from a national competition on structural optimization and additive manufacturing held in Beijing in 2019. The event attracted entries from a large number of experienced aerospace engineers, university researchers, and postgraduate students from across China and abroad. A team of my PhD students (the first three authors of Xiong et al. 2021) entered the competition, equipped with knowledge of topology optimization.

The challenge for each team was to maximize the load-carrying capacity of the design for a given amount of material. Each design must be verified through numerical simulations by the team and then experimentally confirmed by the organizing committee through 3D printing and mechanical testing.

A total of 135 teams entered the preliminary competition, for the load case of a *downward* force. Using the BESO method, my students generated the design shown

3.6 Selecting an Adequate Design Domain

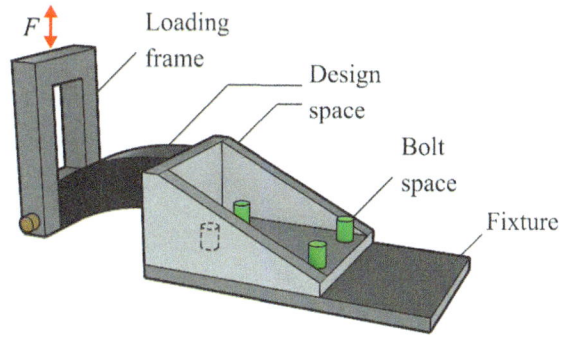

Fig. 3.8 Load and boundary conditions of a hinge arm (reprinted from Lee and Xie 2021, with permission from Elsevier)

in Fig. 3.9, which performed well in the mechanical testing. They qualified for the final competition, along with 17 other teams.

Surprisingly, the only change for the final competition was to reverse the load direction—applying the force *upward*. One might think that, for the simple load and boundary conditions shown in Fig. 3.8, the load direction would have little effect on the design. However, after conducting a few quick numerical simulations using relatively coarse meshes (see Fig. 3.10), my students soon realized that with the upward load, the previously used design domain was inadequate—it should be substantially expanded to leverage the freely available contact support from the base fixture, in front of the bolts (see Fig. 3.8). The large flat areas on the right and top surfaces in Fig. 3.10b clearly indicated that these parts were unintentionally 'cut' by inadequate design domains in Fig. 3.10a.

With the insight gained from the above numerical simulations, my students went on to produce the final optimized design using a fine mesh with an expanded design domain. The result, shown in Fig. 3.11, is distinctly different from the earlier design in Fig. 3.9 for the downward load. During the actual mechanical loading test in the final competition, this design carried the highest load among all entries and my students won the first prize.

Next, we demonstrate that a design domain commonly used for stiffness optimization may not be adequate for optimization with stress constraint. Figure 3.12a shows a classic L-shaped structure, fixed at the top edge and subjected to a uniformly

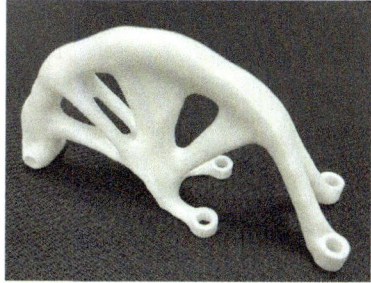

Fig. 3.9 Optimized hinge arm for the load case of a downward force (Photo credit: Yi Min Xie)

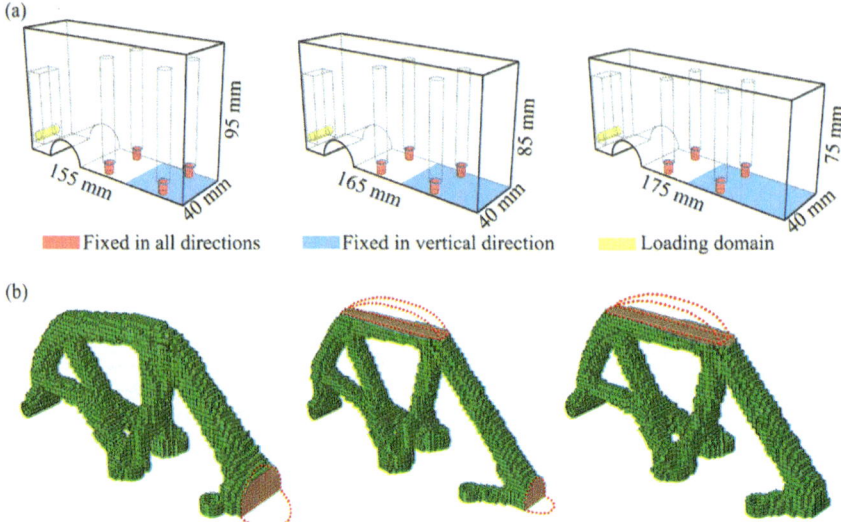

Fig. 3.10 Topology optimization of the hinge arm for the load case of an upward force: **a** initial design domains; **b** optimized designs derived from the design domains in (a), revealing their inadequacies (reprinted from Xiong et al. 2021, licensed under CC-BY 4.0)

Fig. 3.11 Final optimized hinge arm for the load case of an upward load: **a** finite element model; **b** smoothed model; **c** 3D printed prototype after mechanical loading test (reprinted from Xiong et al. 2021, licensed under CC-BY 4.0)

distributed load at the top right corner. The L-shaped design domain is discretized into plane stress quadrilateral elements, with an element size of 1 mm. A typical result for stiff optimization is given in Fig. 3.12b.

Over the past decade, when researchers develop topology optimization techniques for problems with stress constraint, the L-shaped structure has been used by many as a benchmark example, invariably adopting the same design domain as used in stiffness optimization (e.g., Lian et al. 2017; Xia et al. 2018; Fan et al. 2024). A typical result for stress optimization is shown in Fig. 3.12c. This solution is obtained by minimizing the structural volume subject to a constraint on the maximum von Mises stress.

3.6 Selecting an Adequate Design Domain

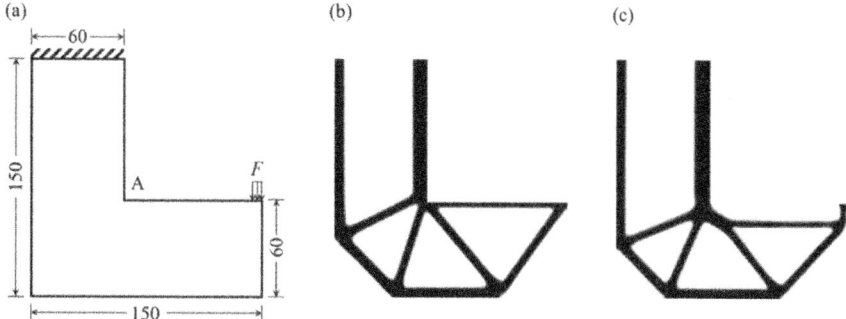

Fig. 3.12 Topology optimization of a L-shaped structure (unit: mm): **a** design domain, and load and support conditions; **b** typical result for stiffness optimization; **c** typical result for stress optimization (by Zicheng Zhuang and Yi Min Xie, printed with permission)

If we look at the design domain in Fig. 3.12a, it is obvious that there is stress concentration at the sharp corner, A. To reduce the stress concentration in this situation, an experienced structural engineer would usually add a fillet to the sharp corner. Inspired by this common practice in structural design, we expand the design domain slightly at the corner by adding a small triangle (see Fig. 3.13a, b). This adjustment results in two new designs shown in Fig. 3.13d, e. Compared to the conventional solution shown in Fig. 3.12c, the final solid volumes of the new solutions have reduced by 10% and 13%, respectively, for the same constraint on the maximum von Mises stress. These substantial savings in material consumption are achieved through very minor expansions of the design domain, by 2% and 6%, respectively. More significantly, expanding the design domain may cost nothing, as void spaces are often freely available in practical cases.

In my view, adopting the original L-shaped design domain for stress optimization is irrational. It brings to mind the Chinese idiom '削足适履', which translates literally to 'cutting the foot to fit the shoe'. A more sensible way to achieve foot comfort would be choosing properly sized and shaped shoes. Similarly, to enable a structure to evolve freely towards its optimal topology, one should select an appropriate design domain to begin with.

For some problems, an 'appropriate' design domain is not so obvious. In such cases, one may either apply the adaptive design domain method (discussed in the following section) or set a very large design domain. Figure 3.13c shows the latter approach of using a large design domain. In this case, the design domain is expanded by 56% in its area (again, at zero cost, as the void space is free), which results in a completely different design in Fig. 3.13f, with its final solid volume reduced by 38% compared to the conventional design shown in Fig. 3.12c for the same stress constraint.

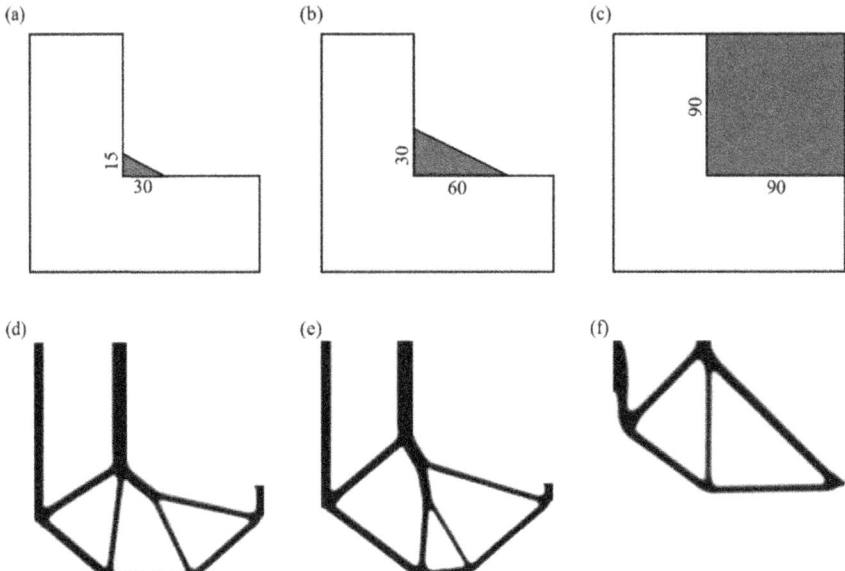

Fig. 3.13 Redefining the design domain for the L-shaped structure: **a–c** expanding the design domain by 2%, 6%, and 56%, respectively; **d–f** optimized designs derived from the design domains in (a), (b), and (c), resulting in 10%, 13%, and 38% reductions in the final solid volume, respectively (by Zicheng Zhuang and Yi Min Xie, printed with permission)

3.7 Using an Adaptive Design Domain

Despite the success in setting an appropriate design domain for the design competition discussed in the previous section, it is not always possible to choose the 'right' dimensions for the design domain. On the other hand, prescribing an oversized design space in the optimization model (e.g., Fig. 3.13c) may significantly increase the computational cost, particularly for 3D problems with a fine mesh of elements. To address these challenges, we have developed an efficient topology optimization method using an *adaptive design domain*, which allows the design space to evolve automatically and intelligently according to structural or functional requirements (Rong et al. 2022).

In the adaptive design domain method, first an initial design domain of certain dimensions is specified, often a cuboid for simplicity. The design domain is divided into a number of cells, and each cell is subdivided into multiple elements (see Fig. 3.14a). At the end of each optimization cycle—which consists of multiple iteration steps—new cells are added in the vicinity of those with relatively high sensitivities, while cells with relatively low sensitivity values may be removed from the current design domain (see Fig. 3.14b). This process of adding and deleting cells is similar to the BESO concept of adding and deleting elements (Huang and Xie 2010). To avoid an abrupt change in the structural topology, the newly added cells are filled

3.7 Using an Adaptive Design Domain

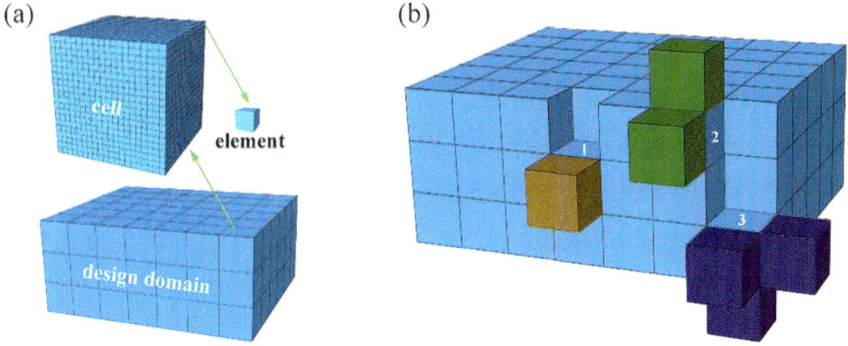

Fig. 3.14 Adaptive design domain: **a** dividing a design domain into cells and then subdividing each cell into elements; **b** cells being added to or deleted from the current design domain (reprinted from Rong et al. 2022, with permission from Elsevier)

Fig. 3.15 Design of a deep beam: **a** load and support conditions; **b** conventional solution for a fixed design domain; **c** new solution from using an adaptive design domain (reprinted from Rong et al. 2022, with permission from Elsevier)

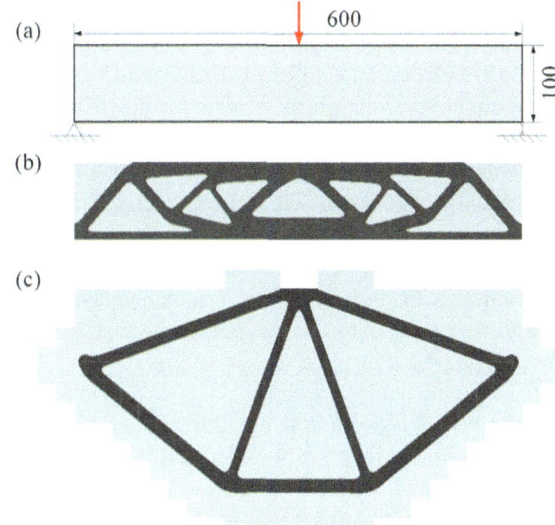

with void elements at the beginning of each cycle. Some of these void elements are gradually switched to solid if they are in the vicinity of existing solid elements with high sensitivity values.

We use a simple example to demonstrate the significant impact of having an adaptive design domain on the design outcome. Figure 3.15a shows a classic design problem of a deep beam that has been solved by numerous researchers, assuming customarily that the rectangular area outlining the locations of the load and supports is the default design domain. For this fixed design domain, a typical solution from topology optimization is shown in Fig. 3.15b. However, if we allow the design domain to expand on all four sides of the rectangular area except for the regions where the

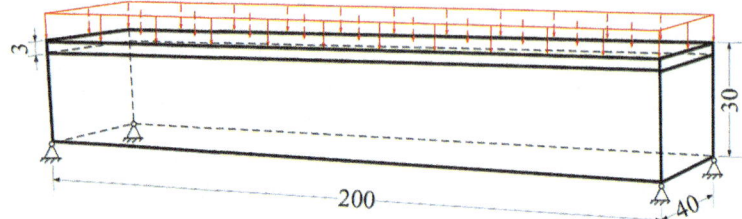

Fig. 3.16 Load and support conditions for a bridge design (reprinted from Rong et al. 2022, with permission from Elsevier)

force is applied, we obtain a totally different solution (see Fig. 3.15c). The new design is not only much simpler in geometry but also far more efficient structurally—for the same amount of material, the compliance of the new design has reduced by 67.7% compared to the previous solution.

Next, we consider a bridge design problem shown in Fig. 3.16, which is very similar to the one in Fig. 3.4a. The top layer is designated as non-design domain for the bridge deck. A uniformly distributed load is applied to the top surface of the initial cuboid design domain, which is divided into $10 \times 2 \times 3$ cuboid cells and each cell is subdivided into $20 \times 20 \times 10$ cubic elements, resulting in a total of 240,000 elements. New cells are allowed to be added anywhere but not below the bottom of the initial design domain. The target volume of each optimized design is set to 20% of that of the initial design domain.

If the initial design domain is kept unchanged during the optimization process, we obtain the result shown in Fig. 3.17a, which closely resembles the previous solution in Fig. 3.4b. Applying the adaptive design domain method yields a series of new designs, of which three are shown in Fig. 3.17b–d. Compared to the initial design in Fig. 3.17a, the compliance of the new design in Fig. 3.17d has reduced by 33%.

Figure 3.18 illustrate the evolution histories of the volume and the compliance, both normalized by their initial values. In addition to the initial design domain, an intermediate and the final design domains are shown in Fig. 3.18. The final design domain contains 688,000 elements.

We have also solved this bridge design problem using a very large, fixed design domain Ω_L (see Fig. 3.19a). Ω_L is divided into 1.92 million elements and its volume is eight times that of the previous initial design domain. Although the height and some details of the new design in Fig. 3.19b differ from those of the previous design in Fig. 3.17d, their compliance values vary by less than 0.3%.

It should be noted that the bridge optimization using the adaptive design domain method was performed on an ordinary personal computer with 16 GB of memory. However, the same computer was out of memory when running the model with the very large design domain shown in Fig. 3.19a. Consequently, we had to use a high-performance workstation to complete the optimization. This example demonstrates that the adaptive design domain method can substantially reduce the computational

3.7 Using an Adaptive Design Domain

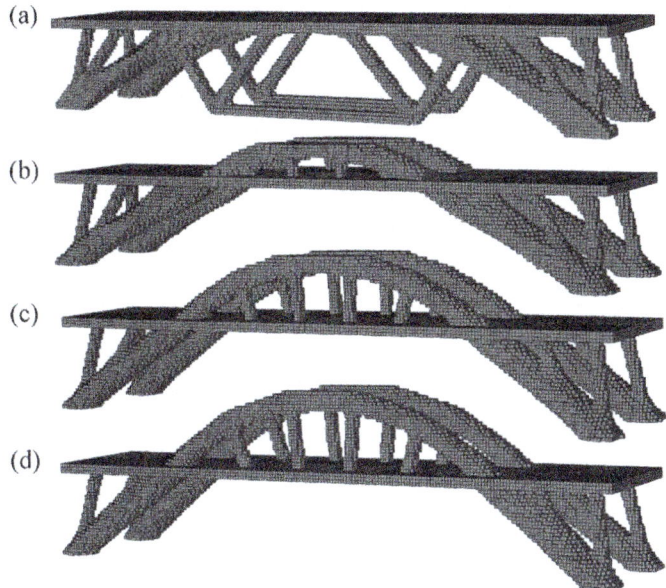

Fig. 3.17 Optimized topologies of the bridge: **a** from the initial, fixed design domain; **b–d** using an adaptive design domain (reprinted from Rong et al. 2022, with permission from Elsevier)

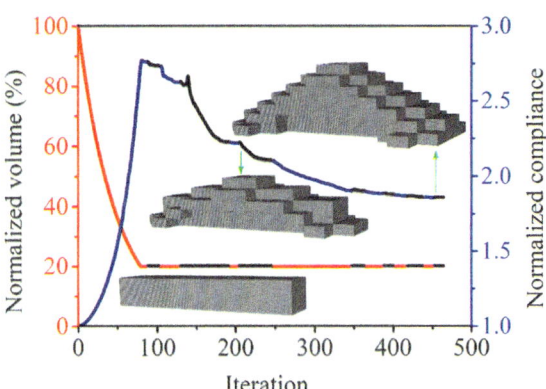

Fig. 3.18 Evolution histories of the normalized volume and compliance, together with the initial, intermediate, and final design domains (reprinted from Rong et al. 2022, with permission from Elsevier)

cost. Further, the computational efficiency can be greatly enhanced by a subdomain-based parallel processing strategy for dealing with multiple cells (Zhao et al. 2023).

In addition to enhancing structural performance, the adaptive design domain method can be used to solve transdisciplinary problems, such as simulating the morphogenesis of growing plant roots. Plants generally exhibit two main types of root systems: taproot and fibrous root. Figure 3.20 shows the growth histories of a taproot and a fibrous root, obtained from a topology optimization method using an adaptive design domain (Rong et al. 2023). Roots grow in soil and around obstacles

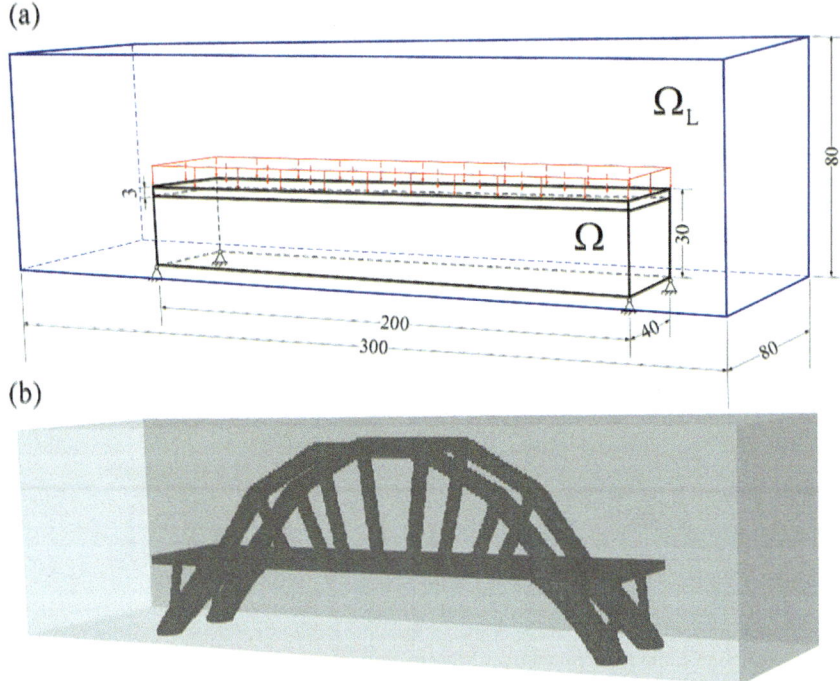

Fig. 3.19 Solving the bridge design problem using a very large, predefined design domain: **a** design domain; **b** optimized design (reprinted from Rong et al. 2022, with permission from Elsevier)

(e.g., rocks) primarily to maximize water and nutrient absorption, a process that can be simulated by the topology optimization method. As the root grows from a tiny rootlet to a mature network occupying a large area, the optimization process can start with a very small initial design domain and then gradually add new cells where the growth is occurring.

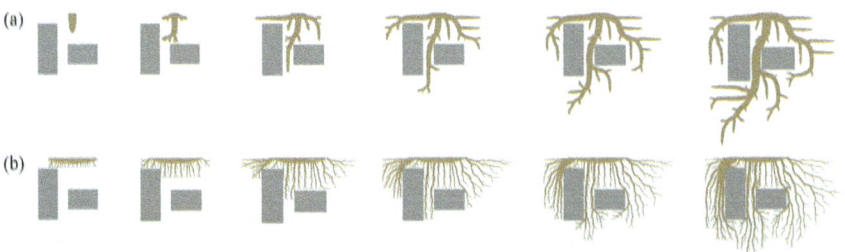

Fig. 3.20 Computational morphogenesis of plant roots growing around rectangular obstacles: **a** a taproot; **b** a fibrous root (reprinted from Rong et al. 2023, with permission from Elsevier)

3.8 Introducing Gaps Within the Design Domain

It should be noted that the root growth process is highly history-dependent and involves a dynamically changing boundary—the interface between root and soil. Consequently, this problem cannot be solved in a single optimization cycle using a very large, fixed design domain. In this case, the adaptive design domain method can be effectively and efficiently used to determine various shapes of the root at different stages of its growth history, as illustrated in Fig. 3.20.

3.8 Introducing Gaps Within the Design Domain

This section presents a novel structure that we have created using topology optimization by introducing a gap in the design domain. Figure 3.21 illustrates a common example of compliant mechanism design. The structure is fixed on the left-hand side at the top and bottom ends (see Fig. 3.21a). When a force is applied at the input port on the left, an inverter mechanism can generate a displacement in the opposite direction at the output port on the right. Such an inverter mechanism can be easily designed using conventional topology optimization by maximizing the output displacement in the desired direction. Figure 3.21b shows a typical design of an inverter mechanism (Ansola Loyola et al. 2018; da Silva et al. 2020).

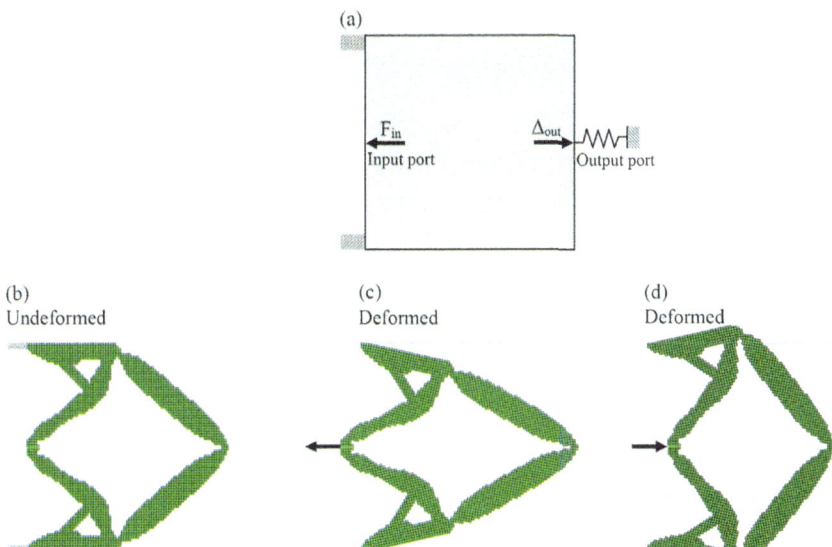

Fig. 3.21 Conventional design of an inverter mechanism: **a** boundary and load conditions; **b** optimized design; **c** a leftward force at the input port generating a rightward displacement at the output port; **d** reversing the direction of the input force resulting in the direction of the output displacement being reversed as well, exhibiting reciprocity (reprinted from Shobeiri and Xie 2025, licensed under CC-BY 4.0)

In most physical systems, the law of reciprocity ensures that a structure responds to an input force in a 'symmetric' manner when the force direction is reversed (Wang et al. 2023), as illustrated in Fig. 3.21c, d—when the direction of the input force is reversed, the output port moves in the opposite direction. Such reciprocal behaviour is commonly observed in most engineering structures.

In the past decade, there has been growing interest in developing non-reciprocal structures and metamaterials that respond differently when the force direction is reversed. Existing research in this area is primarily based on the concept of force-direction-dependent buckling of slender members in a structure (Coulais et al. 2017) or nanofillers embedded in a composite material (Wang et al. 2023).

We have recently developed a method of using gaps or contact surfaces in a continuum structure to achieve mechanical non-reciprocity. Contrary to slender members that are weak under compression due to buckling, the gaps act as weak points under tension but maintain the stiffness and strength of an unbroken material when subjected to compression. This simple yet general concept has enabled us to create a series of non-reciprocal structures and metamaterials, one of which is illustrated in Figs. 3.22 and 3.23 (Shobeiri and Xie 2025).

Here, we redesign the inverter mechanism by introducing a vertical gap of zero thickness in the middle of the design domain (see Fig. 3.22a). Topology optimization is performed with two load cases: one for the input force to act to the left and the other to the right. For both load cases, the design objective is to maximize the output

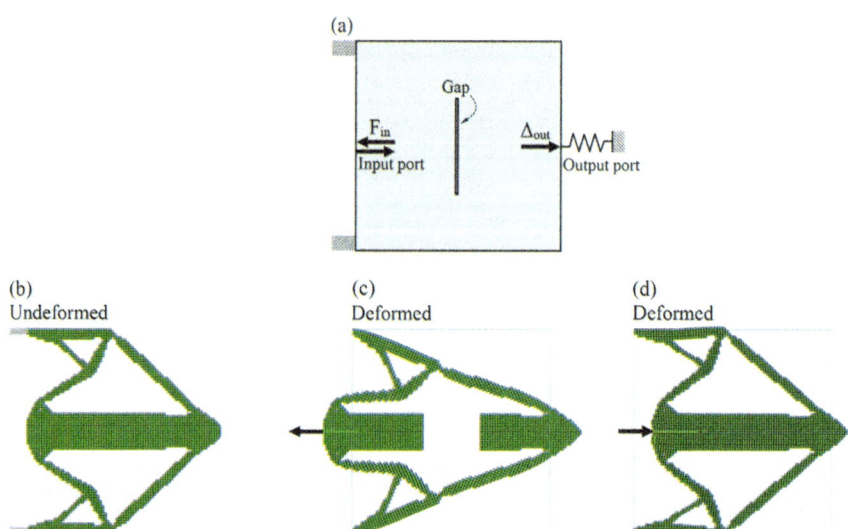

Fig. 3.22 Novel design of a non-reciprocal structure: **a** introducing a gap within the design domain; **b** optimized design (with an invisible gap in the horizontal bar); **c** a leftward force at the input port generating a rightward displacement at the output port; **d** reversing the direction of the input force without changing the direction of the output displacement, exhibiting non-reciprocity (reprinted from Shobeiri and Xie 2025, licensed under CC-BY 4.0)

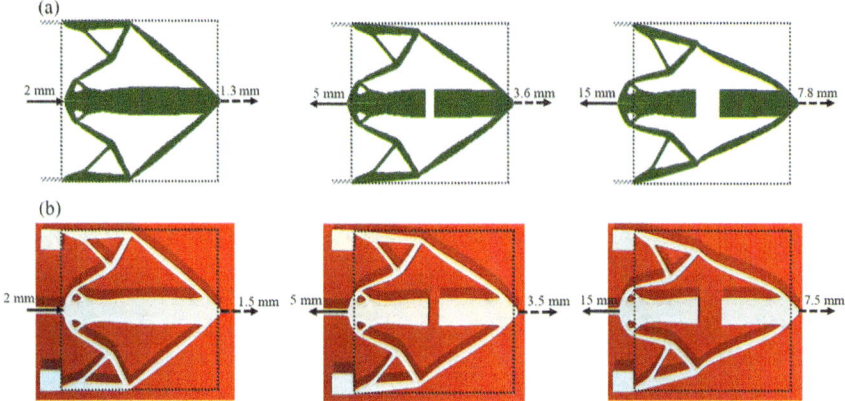

Fig. 3.23 Comparison between numerical simulations and experimental observations of non-reciprocal responses: **a** finite element analysis results; **b** experimental results (reprinted from Shobeiri and Xie 2025, licensed under CC-BY 4.0)

displacement in one direction—to the right. By taking a weighted average of the sensitivities from the two load cases, we achieve a novel design shown in Fig. 3.22b, which includes an invisible vertical gap in the middle of the horizontal bar.

Unlike the conventional design shown in Fig. 3.21b, the new structure is non-reciprocal: regardless of whether the input force acts to the left or right, the output port always moves to the right! The vertical gap introduced into the design domain has played a critical role in producing this unusual behaviour. When the input force acts to the left, the structure deforms in the same manner as the conventional inverter since the horizontal bar broken in the middle has little effect. However, when the input force acts to the right, the gap closes and the horizontal bar behaves as if it were continuous, effectively pushing the output port to the right.

To verify our design from topology optimization, we have conducted experiments on 3D printed prototypes and performed finite element analysis considering geometrical non-linearity due to large deformation. Figure 3.23 shows that the results from numerical simulations agree well with experimental observations.

This example demonstrates that by creatively redefining the design domain, we can unlock new possibilities for achieving extraordinary structural designs.

3.9 Conclusion

Specifying the design domain for topology optimization seems to be a trivial and obvious task. However, through a wide range of examples, we have demonstrated that redefining the design domain provides perhaps the greatest opportunities for creativity and innovation in structural design. By strategically exploring different design domains, we can quickly find a wide variety of geometrically distinct and structurally efficient designs. In practice, my team often uses this approach when working with architects or clients during the early stages of a design project. This enables us to provide them with multiple options and maximize our chances of continued collaboration. In Chap. 7, we show several large-scale, practical projects where this simple yet effective approach has been successfully applied.

References

Ansola Loyola, R., Querin, O. M., Garaigordobil Jiménez, A. and Alonso Gordoa, C. (2018) A sequential element rejection and admission (SERA) topology optimization code written in Matlab. *Struct. Multidisc. Optim.* **58**, 1297–1310.
Bonner, J. (2017) *Islamic Geometric Patterns: Their Historical Development and Traditional Methods of Construction.* New York: Springer.
Clausen, A., Aage, N. and Sigmund, O. (2014) Topology optimization with flexible void area. *Struct. Multidisc. Optim.* **50**, 927–943.
Coulais, C., Sounas, D. and Alù, A. (2017) Static non-reciprocity in mechanical metamaterials. *Nature* **542**, 461–464.
da Silva, G. A., Beck, A. T. and Sigmund, O. (2020) Topology optimization of compliant mechanisms considering stress constraints, manufacturing uncertainty and geometric nonlinearity. *Comput. Methods Appl. Mech. Eng.* **365**, 112972.
Fan, Z., Gao, L. and Li H. (2024) Isogeometric topology optimization method for design with local stress constraints. *Comput. Struct.* **305**, 107564.
He, Y. and Xie, Y. M. (2024) Clustering-based topology optimization of periodic structures with variable orientations of unit cells. *Eng. Struct.* **316**, 118518.
Huang, X. and Xie, Y. M. (2008) Optimal design of periodic structures using evolutionary topology optimization. *Struct. Multidisc. Optim.* **36**, 597–606.
Huang, X. and Xie, Y. M. (2010) *Evolutionary Topology Optimization of Continuum Structures: Methods and Applications.* Chichester: John Wiley & Sons.
Lee, T.-U. and Xie, Y. M. (2021) Simultaneously optimizing supports and topology in structural design. *Finite Elem. Anal. Des.* **197**, 103633.
Li, H., Kondoh, T., Jolivet, P., Furuta, K., Yamada, T., Zhu, B., Izui, K. and Nishiwaki, S. (2022) Three-dimensional topology optimization of a fluid–structure system using body-fitted mesh adaption based on the level-set method. *Appl. Math. Model.* **101**, 276–308.
Li, Y. and Xie, Y. M. (2021a) Evolutionary topology optimization for structures made of multiple materials with different properties in tension and compression. *Compos. Struct.* **259**, 113497.
Li, Y. and Xie, Y. M. (2021b) Evolutionary topology optimization of spatial steel–concrete structures. *J. Int. Assoc. Shell Spat. Struct.* **62**, 102–112.
Lian, H., Christiansen, A. N., Tortorelli, D. A., Sigmund, O. and Aage, N. (2017) Combined shape and topology optimization for minimization of maximal von Mises stress. *Struct. Multidisc. Optim.* **55**, 1541–1557.
Meng, X., Zhang, L.-Y., Zhao, Z.-L. and Xie, Y. M. (2023) A direct approach to achieving efficient free-form shells with embedded geometrical patterns. *Thin-Walled Struct.* **185**, 110559.
Nishizawa R. (2010) Teshima Art Museum. *El Croquis* **155**, 192–205.
Rong, Y., Zhao, Z.-L., Feng, X.-Q. and Xie, Y.M. (2022) Structural topology optimization with an adaptive design domain. *Comput. Methods Appl. Mech. Eng.* **389**, 114382.
Rong, Y., Zhao, Z.-L., Feng, X.-Q., Yang, J. and Xie, Y. M. (2023) Computational morphomechanics of growing plant roots. *J. Mech. Phys. Solids* **178**, 105346.
Shobeiri V. and Xie, Y. M. (2025) Topology optimization for creating nonreciprocal compliant mechanisms: Numerical and experimental investigations. *Extreme Mech. Lett.* **77**, 102345.

References

Wang, X., Li, Z., Wang, S., Sano, K., Sun, Z., Shao, Z., Takeishi, A., Matsubara, S., Okumura, D., Sakai, N., Sasaki, T., Aida, T. and Ishida, Y. (2023) Mechanical nonreciprocity in a uniform composite material. *Science* **380**, 192–198.

Xia, L., Zhang, L., Xia, Q. and Shi, T. (2018) Stress-based topology optimization using bi-directional evolutionary structural optimization method. *Comput. Methods Appl. Mech. Eng.* **333**, 356–370.

XIE Technologies (2024) Ameba: Topology optimization software based on BESO. https://ameba.xieym.com. Accessed 8 December 2024.

Xie, Y. M. (2022) Generalized topology optimization for architectural design. *Archit. Intell.* **1**, 2.

Xiong, Y., Bao, D., Yan, X., Xu, T. and Xie, Y. M. (2021) Lessons learnt from a national competition on structural optimization and additive manufacturing. *Curr. Chin. Sci.* **1**, 151–159.

Yang, K., Zhao, Z.-L., He, Y., Zhou, S., Zhou, Q., Huang, W. and Xie, Y. M. (2019) Simple and effective strategies for achieving diverse and competitive structural designs. *Extreme Mech. Lett.* **30**, 100481.

Zhang, W. and Zhou, Y. (2020) *The Feature-driven Method for Structural Optimization.* Amsterdam: Elsevier.

Zhao, Z.-L., Rong, Y., Yan, Y., Feng, X.-Q. and Xie, Y. M. (2023) A subdomain-based parallel strategy for structural topology optimization. *Acta Mech. Sin.* **39**, 422357.

Zhou, Q., Shen, W., Wang, J., Zhou, Y. Y. and Xie, Y. M. (2018) Ameba: A new topology optimization tool for architectural design. *Proc. IASS Annu. Symp.*, Boston, 16–20 July 2018.

Open Access This chapter is licensed under the terms of the Creative Commons Attribution 4.0 International License (http://creativecommons.org/licenses/by/4.0/), which permits use, sharing, adaptation, distribution and reproduction in any medium or format, as long as you give appropriate credit to the original author(s) and the source, provide a link to the Creative Commons license and indicate if changes were made.

The images or other third party material in this chapter are included in the chapter's Creative Commons license, unless indicated otherwise in a credit line to the material. If material is not included in the chapter's Creative Commons license and your intended use is not permitted by statutory regulation or exceeds the permitted use, you will need to obtain permission directly from the copyright holder.

Chapter 4
Optimizing Support Locations

Structural topology optimization is usually performed with predetermined support conditions. In this chapter, we present a general method for optimizing support locations and show how it can be integrated with topology optimization. By allowing the optimizer to simultaneously determine support locations and structural topology, we unlock new opportunities to enhance structural performance and achieve innovative and efficient designs.

4.1 Introduction

Much of the previous research on optimizing support locations has focused on the layout of columns supporting a plate structure. Typically, the plate represents a roof or a concrete floor. Studies have shown that the layout of columns significantly affects the local deformation of the plate and the overall stiffness of the plate–column system (Jang et al. 2009; Meng et al. 2021; Zelickman and Amir 2022a, b).

Figure 4.1 shows an example of a bean-shaped flat roof supported by vertical columns (Meng et al. 2021). Using an optimality criteria method, we can optimize the column locations to maximize the overall stiffness of the structure (which is equivalent to minimizing the compliance, C). In this example, we set the number of final columns to seven. The optimization process starts from 1159 candidate columns within an allowable area and gradually finds the final seven columns (see Fig. 4.1a). The optimized design is illustrated in Fig. 4.1b. The final column locations meet both engineering and architectural requirements: the structural performance is optimized and the number of columns satisfies the designer's prescribed condition. To verify the optimization result, a set of arbitrary, manually arranged columns is tested, as shown in Fig. 4.1c. It can be seen that the optimized design exhibits much lower compliance than any of the manual designs.

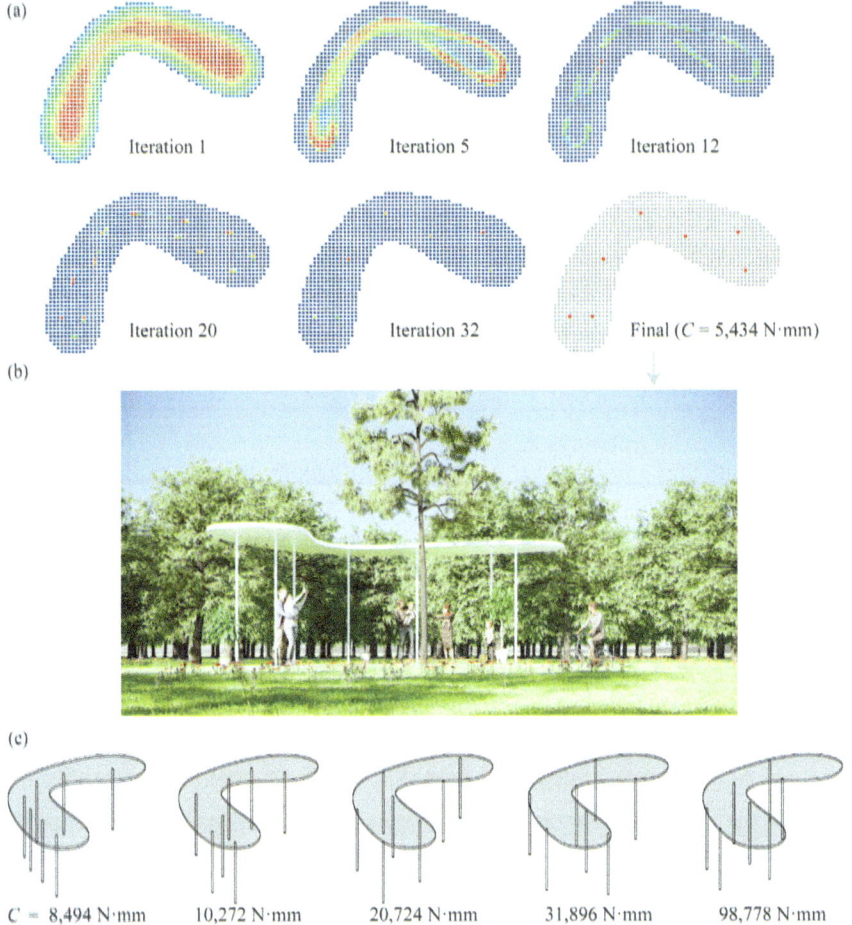

Fig. 4.1 Optimizing column locations under a bean-shaped flat roof: **a** optimization process; **b** rendering of the optimized design; **c** five manual designs and their compliance values (reprinted from Meng et al. 2021, licensed under CC-BY 4.0)

However, it should be noted that the structural performance of the seemingly simple roof–column system is extremely sensitive to the column locations, and the optimization algorithms can often converge to local minima that are far from the true optimum. To mitigate this issue, Meng et al. (2021) and Zelickman and Amir (2022b) proposed several techniques to avoid local optima. However, these techniques seem to be ad hoc and require further refinement to enhance their reliability and robustness.

For dynamic problems, Jihong and Weihong (2006) developed an optimization technique that employs elastic springs to represent supports. By constraining the cost of springs, optimized support locations can be determined to maximize the fundamental natural frequency of the structure. In their work, only support locations are optimized and the structural topology remains unchanged.

Buhl (2002) conducted the first study on the simultaneous optimization of support locations and structural topology. Linear elastic springs are introduced in a finite element model at nodes where supports are permitted. As a result, the main diagonal of the global stiffness matrix is modified by adding the stiffness contributions from the springs. The stiffness of the springs can vary continuously between the lower and upper bounds, representing either the absence or presence of supports. By constraining the number (or the total cost) of the supports, support locations and structural topology can be optimized concurrently.

In this chapter, we present an alternative method for the simultaneous optimization of support locations and structural topology. Unlike previous approaches, our method does not require introducing springs to the finite element model. Instead, we add a layer of support elements with assigned boundary conditions to the structural design domain. This method can be easily implemented and conveniently integrated with widely used commercial finite element analysis (FEA) software such as Abaqus and Ansys, without needing to access or modify the code for the global stiffness matrix or any other parts of the FEA code. The effectiveness of this method is demonstrated through a series of numerical examples, including applications to two complex 3D design problems. Much of the material presented in this chapter is based on the work of Lee and Xie (2021).

4.2 Optimizing Support Locations and Structural Topology

4.2.1 Problem Definition and Statement

In our method for optimizing support locations, we add a thin layer of support domain to the structural design domain, with the boundary condition assigned to the support domain (see Fig. 4.2a). During the meshing process, the support domain is divided into a single layer of elements (see Fig. 4.2b). Elements within the support domain and the structural design domain are referred to as support and structural elements, respectively.

The optimization problem can be formulated to minimize the compliance, C, subject to constraints on the cost of supports and the volume of the structure. Mathematically, this problem can be expressed as

$$\text{Minimize}: C = \mathbf{f}^T \mathbf{u} \tag{4.1a}$$

$$\text{Subject to}: \mathbf{K}\mathbf{u} = \mathbf{f} \tag{4.1b}$$

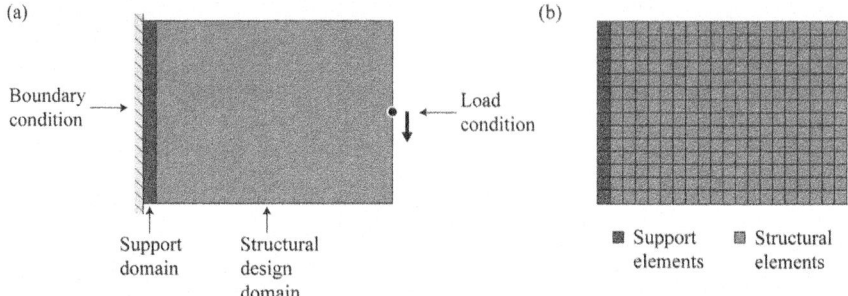

Fig. 4.2 Initial setup for the simultaneous optimization of support locations and topology of a short cantilever: **a** adding a thin layer of support domain to the structural design domain; **b** dividing the support domain into a single layer of elements (reprinted from Lee and Xie 2021, with permission from Elsevier)

$$: \sum_{i=1}^{N} S_i y_i = S^* \qquad (4.1c)$$

$$: 0 < y_{\min} \leq y_i \leq 1 \qquad (4.1d)$$

$$: \sum_{i=1}^{M} V_i x_i = V^* \qquad (4.1e)$$

$$: x_i = x_{\min} \text{ or } 1 \qquad (4.1f)$$

where $\mathbf{f}$ and $\mathbf{u}$ are the nodal force and displacement vectors, respectively. Equation (4.1b) ensures the static equilibrium of the structural system, where $\mathbf{K}$ is the global stiffness matrix. Equation (4.1c) is the cost constraint on the supports, where S_i is the cost of the ith support element, S^* is the prescribed total cost, and N is the total number of elements in the support domain. Equation (4.1e) is the volume constraint on the structural elements, where V_i is the volume of the ith element, V^* is the prescribed target volume of the structure, and M is the total number of elements in the structural model. y_i and x_i are the ith design variables denoting densities of the support and structural elements, respectively. $y_{\min}$ and $x_{\min}$ are small values (e.g., 0.001), representing the absence of the support and structural elements, respectively.

For most cases, the cost constraint on supports (Eq. 4.1c) can be simplified using a more straightforward description:

$$\sum_{i=1}^{N} y_i = N^* \qquad (4.2)$$

where N^* is the prescribed target number of support elements (or locations) in the final design.

4.2.2 Optimization of Support Locations

To consider all candidate support locations during the optimization process, the continuous design variable of every support element, y_i, is updated using a standard optimality criteria method (Bendsøe 1995; Sigmund 2001), as follows

$$y_i^{\text{new}} = \begin{cases} \max\{y_{\min}, y_i^k(1-m)\}, \text{ if } y_i^k B_i^\eta \leq \max\{y_{\min}, y_i^k(1-m)\} \\ y_i^k B_i^\eta, \text{ if } \max\{y_{\min}, y_i^k(1-m)\} < y_i^k B_i^\eta < \min\{1, y_i^k(1+m)\} \\ \min\{1, y_i^k(1+m)\}, \text{ if } y_i^k B_i^\eta \geq \min\{1, y_i^k(1+m)\} \end{cases} \quad (4.3)$$

where y_i^k denotes the ith design variable of the support domain at the kth iteration, m is the move limit (typically $m = 0.2$), η is a numerical damping coefficient (usually $\eta = 0.5$), and B_i is determined by the optimality condition based on the elastic strain energy density, U_i, of the ith support element. The entire B_i^η term is defined as

$$B_i^\eta = \lambda \sqrt{U_i} \quad (4.4)$$

where λ is a Lagrange multiplier that can be determined using a bisection algorithm (Kumar and Suresh 2021) to enforce the constraint on the support elements stated in Eq. (4.2). In the bisection algorithm

$$\lambda = (b_{lower} + b_{upper})/2 \quad (4.5)$$

where the lower and upper bounds b_{lower} and b_{upper} are initially set as 0 and 10^6, respectively, and their values are updated according to the summation of current design variables, y_i, as follows

$$b_{lower} = \lambda, \text{ if } \sum_{i=1}^{N} y_i < N^* \quad (4.6a)$$

$$b_{upper} = \lambda, \text{ if } \sum_{i=1}^{N} y_i \geq N^* \quad (4.6b)$$

The bisectional process using Eqs. (4.5) and (4.6a, 4.6b) is repeated until $b_{upper} - b_{lower} < 10^{-8}$. This ensures that the constraint on the support elements (Eq. 4.2) is satisfied at the end of each optimization iteration.

The initial value of y_i is set as

$$y_0 = \frac{N^*}{N} \quad (4.7)$$

meaning that the prescribed amount of support material is uniformly distributed to all candidate support locations at the beginning of the optimization process.

Adopting the commonly used solid isotropic material with penalization (SIMP) model (Rozvany et al. 1992), the material property of support elements is interpolated as a function of the element density:

$$E(y_i) = y_i^p E_0 \qquad (4.8)$$

where $E(y_i)$ is the Young's modulus of the ith support element, E_0 is the Young's modulus of the support element when it is solid, and p is the penalty exponent of support elements. The effects of using $p = 1$ and $p = 3$ are discussed in detail in Sect. 4.3.1.

4.2.3 BESO Method for Optimizing Structural Topology

For the structural elements, a material interpolation scheme similar to Eq. (4.8) may be used:

$$E(x_i) = x_i^P E_1 \qquad (4.9)$$

where $E(x_i)$ is the Young's modulus of the ith structural element, E_1 is the Young's modulus of the structural element when it is solid, and P is the penalty exponent of structural elements. In the examples in this chapter, $P = 3$ is used. The design variable of each structural element, x_i, , is updated using the bi-directional evolutionary structural optimization (BESO) method. A comprehensive discussion on the BESO method and its applications has been given by Huang and Xie (2010). Here, we provide a brief introduction.

The BESO method determines the structural topology based on the relative ranking of sensitivity numbers of elements. The sensitivity number of the ith element is defined as

$$\alpha_i = -\frac{1}{V_i P}\frac{\partial C}{\partial x_i} = \frac{1}{V_i P}\mathbf{u}^T \frac{\partial \mathbf{K}}{\partial x_i}\mathbf{u} = x_i^{P-1}\frac{\mathbf{u}_i^T \mathbf{K}_i^1 \mathbf{u}_i}{V_i} = \frac{U_i}{x_i} \qquad (4.10)$$

where $\mathbf{u}_i$ and U_i are the displacement vector and the elastic strain energy density of the ith element in the structural domain, respectively. $\mathbf{K}_i^1$ is the stiffness matrix of the ith element when it is solid. Note that in the soft-kill BESO method, the design variables, x_i, , can only have two discrete values: 1 (for solid elements) or x_{min} (for void elements). When $x_i = 1$, the sensitivity number α_i is simply equal to the strain energy density U_i. For the void elements ($x_i = x_{min}$), the sensitivity number α_i is usually very small because

$$\alpha_i = x_{min}^{P-1}\frac{\mathbf{u}_i^T \mathbf{K}_i^1 \mathbf{u}_i}{V_i}, \text{ when } x_i = x_{min} \qquad (4.11)$$

4.2 Optimizing Support Locations and Structural Topology

To avoid checkerboard patterns and mesh-dependency issues, we first calculate the nodal sensitivity number, α_j^n, by averaging the elemental sensitivity number, α_i, of all elements connected to this node (Huang and Xie 2010). Then, we apply the following filtering to the nodal sensitivity number

$$\hat{\alpha}_i = \frac{\sum_{j=1}^{K} w(r_{ij}) \alpha_j^n}{\sum_{j=1}^{K} w(r_{ij})} \quad (4.12)$$

where $\hat{\alpha}_i$ is the filtered sensitivity number of the ith element, K is the total number of nodes in the sub-domain within a circle (in 2D) or a sphere (in 3D) centred at the ith element and having a prescribed filter radius of $r_{\min}$, and the linear weight factor $\omega(r_{ij})$ is defined as

$$w(r_{ij}) = \max\{0, (r_{\min} - r_{ij})\} \quad (4.13)$$

where r_{ij} is the distance between the centre of the ith element and the jth node.

Further, averaging the sensitivity number with its historical values is an effective way to improve the stability and convergence of the BESO process (Huang and Xie 2007). This is achieved by averaging the filtered sensitivity of the current iteration with that of the previous iteration:

$$\tilde{\alpha}_i = \frac{\hat{\alpha}_i^k + \hat{\alpha}_i^{k-1}}{2}, \quad (k \geq 2) \quad (4.14)$$

where $\tilde{\alpha}_i$ is the averaged sensitivity number. Then let $\hat{\alpha}_i^k = \tilde{\alpha}_i$, which will be used for the next iteration.

Before elements are removed from or added to the current design, the target volume for the next iteration (V^{k+1}) is determined first. Since the target structural volume (V^*) can be greater or smaller than the volume of the initial guess design, the volume of each iteration may decrease or increase step-by-step until the target volume is achieved (i.e., until the volume constraint Eq. (4.1e) is satisfied). The evolution of the structural volume is given by

$$V^{k+1} = V^k(1 \pm ER) \quad (4.15)$$

where ER is the evolutionary rate. Once the target volume is reached, the structural volume will be kept constant for the remaining iterations as

$$V^{k+1} = V^* \quad (4.16)$$

Using the value of V^{k+1} from Eq. (4.15) or (4.16), the number of solid elements in the next iteration (N^{k+1}) can be determined. All elements—both solid and void—are sorted in descending order based on their sensitivity numbers calculated from Eq. (4.14). From this sorted list, the top N^{k+1} elements are set as solid and the remaining elements are set as void. The optimization process then moves to the next iteration.

4.2.4 Convergence Criterion

The iterative process of simultaneous optimization of support locations and structural topology continues until the constraints on the support and structural elements are fulfilled and the following convergence criterion relating to the variation of the objective function (the compliance, C) is satisfied

$$\frac{\left|\sum_{n=1}^{L}(C_{k-n+1} - C_{k-L-n+1})\right|}{\sum_{n=1}^{L} C_{k-n+1}} \leq \tau \quad (4.17)$$

where τ is the allowable convergence tolerance and L is an integer number. Typically, τ is set to 0.01% and L is set to 5, which implies that the change in the compliance over the last 10 iterations becomes acceptably small. Note that the compliance value in this chapter includes contributions from both structural and support elements.

4.2.5 Computational Workflow and Implementation

The computational workflow for the simultaneous optimization of support locations and structural topology is illustrated in Fig. 4.3. This process requires the utilization of three different software tools. First, an initial geometry is prepared using computer-aided design (CAD) software, e.g., Rhino Grasshopper, if the structure has a complex shape, as in the 3D examples presented in Sect. 4.4. Second, an initial finite element model is generated using FEA software, e.g., Abaqus. Third, design optimization is performed using a Python script, typically executed in the Spyder environment (Zuo and Xie 2015).

4.2 Optimizing Support Locations and Structural Topology 67

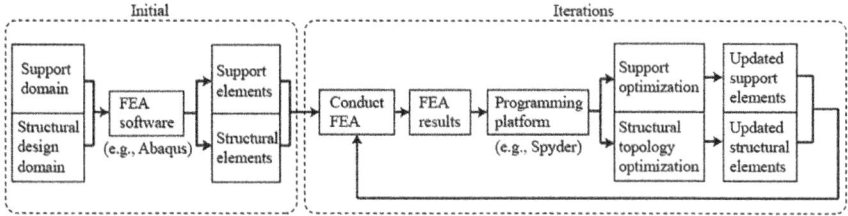

Fig. 4.3 Computational workflow for the simultaneous optimization of support locations and structural topology (reprinted from Lee and Xie 2021, with permission from Elsevier)

4.2.6 Results of a Test Example

The entire optimization process is first tested using a classic short cantilever example (Huang and Xie 2010), as illustrated in Fig. 4.2a. The dimensions of the cantilever are 80 mm × 50 mm for the structural design domain and 1 mm × 50 mm for the support domain. A fixed boundary condition is assigned to the left-hand side of the support domain, and a 100 N force is applied vertically downwards at the centre of the free end. Quadrilateral shell elements are used, with the mesh size and shell thickness both set to 1 mm. Young's modulus $E_0 = E_1 = 100$ GPa and Poisson's ratio $\nu = 0.3$ are assumed. The volume fraction of the final structural topology is set to 50%. From 50 candidate support elements, the target number of supports, N^*, is tested for 1, 2, 4, 5, 6, 8, 10, 14, 28, and 32 to examine the effect of the number of support locations. Other optimization parameters are $p = 1$, $ER = 1\%$, and $r_{\min} = 3$ mm.

Figure 4.4 shows the optimization results. It is observed that the use of different N^* gives different structural topologies, demonstrating that a change in support locations may lead to a significant variation in the structural topology. As expected, larger N^* results in lower compliance, indicating that stiffer structures are generated when using more support elements. It is also noted that the arrangements of support elements exhibit two trends: smaller values of N^* result in 'discrete support locations', while larger values lead to 'continuous supporting regions'. Moreover, when $N^* = 1$, two 'grey' support elements with $y_i = 0.5$ are generated, rather than a single fully solid support element. This is because the short cantilever requires a minimum of two supporting regions due to its symmetrical setting. For the same reason, when $N^* = 5$, two sets of '2.5' support elements are generated.

Further, when N^* is set to a large value, such as 28 or 32, some support elements may become disconnected from the main structure. These disconnected support elements are redundant and have a negligible contribution to the final compliance value.

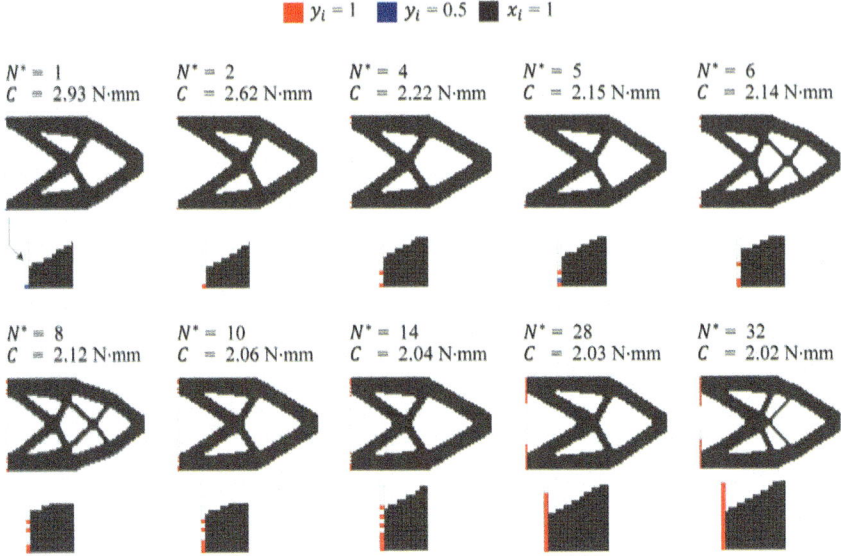

Fig. 4.4 Optimization results of a short cantilever, with each sub-figure showing the number of supports, final compliance, support locations, and structural topology (reprinted from Lee and Xie 2021, with permission from Elsevier)

4.2.7 Verification of Optimization Result

The optimization result for $N^* = 4$ from Fig. 4.4 is re-examined, through a support location analysis, to validate the present simultaneous optimization method. For the final four support elements, it is reasonable to expect two of them to be located at the very top and bottom ends, to maximize resistance to the overall bending moment created by the point load on the other side of the cantilever. However, the optimal locations of the two inner support elements are not obvious. To verify the optimized solution, we manually relocate these two inner support elements towards or away from each other, corresponding to negative and positive moves, respectively, as illustrated in Fig. 4.5a. Note that while the inner support elements are moved, the structure itself remains unchanged.

Figure 4.5b compares the compliance values of the structural system when the move is equal to $-3, -2, -1, 0, 1$, and 2. The optimized design (move = 0) exhibits lower compliance than all other designs generated from positive and negative moves. This comparison confirms that the simultaneous optimization method has successfully found the best support locations from many possible combinations. In this case, the optimized support elements can effectively resist rotations in each of the two supported areas by separating the two support elements and maintaining a certain distance between them.

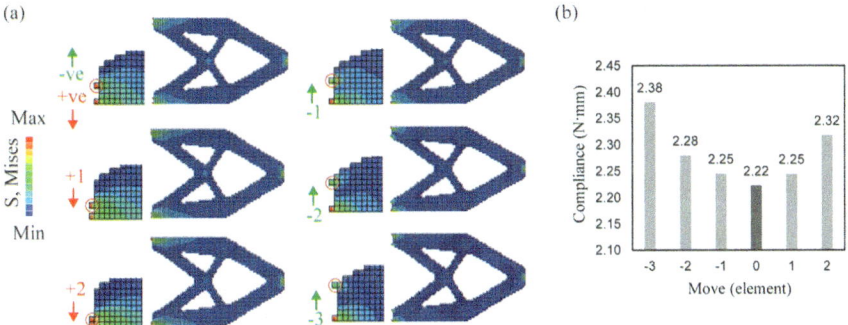

Fig. 4.5 Analysis of support locations: **a** relocation of supports; **b** compliance comparison (reprinted from Lee and Xie 2021, with permission from Elsevier)

4.3 Discussion

4.3.1 Material Model of Support Elements

The material model of support elements is defined by Eq. (4.8). Although $p = 3$ is usually adopted in the SIMP interpolation model, we have deliberately used $p = 1$ for the short cantilever example in Sect. 4.2.6. Such a linear material model is called 'the ersatz material model' (Allaire et al. 2004). The linear material model has been shown to produce better solutions, sometimes, for certain problems in topology optimization (Hu et al. 2020; Huang 2021).

To demonstrate the effect of using different penalty exponents for the support elements, we re-optimize the previous short cantilever using $p = 3$. Figure 4.6 shows the results of support locations, structural topology and evolutionary histories of compliance and volume fraction for cases of $p = 1$ and $p = 3$, with $N^* = 4$ and $N^* = 5$. The comparison reveals that using $p = 1$ and $p = 3$ may result in similar structural topologies but different support locations. The generation of different support locations is primarily due to different optimization histories and tendencies, as shown in Fig. 4.6c, d. When the penalty exponent p is set to 3, the Young's modulus of the support elements is reduced excessively in early iterations, causing large deformations in support elements and an over-estimation of their strain energy (see Fig. 4.6d). Besides, using $p = 3$ leads to a rapid formation of support locations and 'clustering' of support elements (see Fig. 4.6b). This rapid formation occurs because the densities of some of the support elements quickly converge to 1 in the first few iterations and remain unchanged in later iterations. These elements then become 'fixed' support locations in later iterations while the structural topology is optimized.

Figure 4.6a, b shows that $p = 3$ results in larger final compliance values than $p = 1$, indicating $p = 1$ produces better designs. When $p = 1$, separate support locations emerge in later iterations, as shown in Fig. 4.6a. Using the linear material

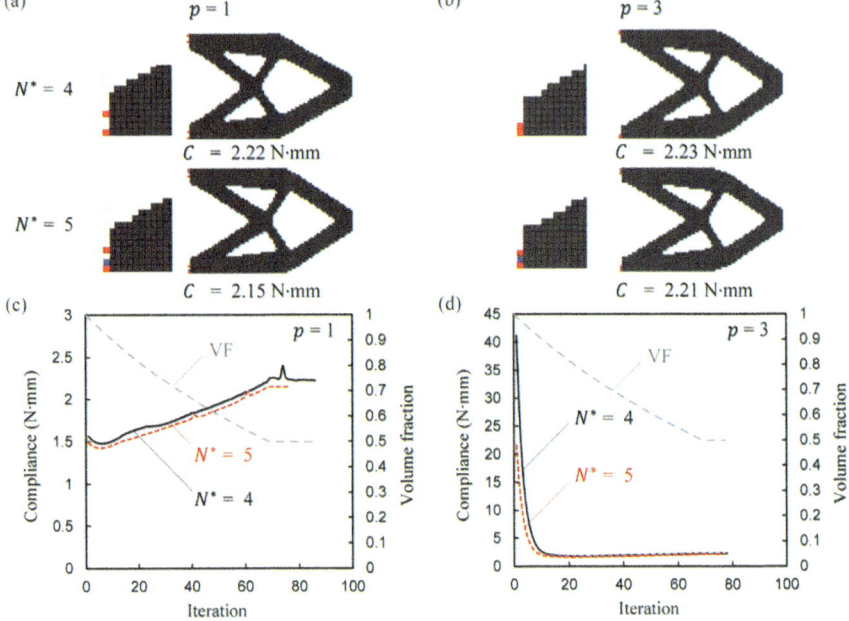

Fig. 4.6 Effect of using different penalty exponents for support elements: **a** and **b** final support locations, compliance, and structural topology; **c** and **d** evolutionary histories of compliance and volume fraction (VF). $p = 1$ for (a) and (c), and $p = 3$ for (b) and (d) (reprinted from Lee and Xie 2021, with permission from Elsevier)

model allows the support locations to form slowly, which benefits the simultaneous optimization method.

Nevertheless, when $p = 1$ is used, there is a reasonable concern about possibly generating unwanted grey support elements whose densities are between 0 and 1. Although we cannot provide a theoretical justification, our numerical tests on a large number of examples seem to indicate that grey support elements rarely remain in the final optimized designs when N^* is small. As mentioned in Sect. 4.2.6, grey support elements with $y_i = 0.5$ sometimes appear in symmetrical structures when the prescribed number for N^* makes it impossible to maintain the symmetry of the structural system if all the support elements are fully solid (see Fig. 4.4, $N^* = 1$ and $N^* = 5$).

To further corroborate the observations made above, three additional examples are tested, including a two-bar frame (Xie and Steven 1993), a Michell-type structure (Xie and Steven 1993), and a 2D bridge (Buhl 2002), as shown in Fig. 4.7. The results again demonstrate a strong clustering effect of support elements when $p = 3$. This is most evident in the bridge example, where separate piers are formed when $p = 1$, while a joined pier is generated when $p = 3$. It is seen that using $p = 1$ consistently results in better designs with lower compliance values. However, in one case (see Fig. 4.7d, $N^* = 4$), $p = 1$ leads to a slightly higher compliance. Significantly different topologies are obtained with $p = 1$ and $p = 3$ for both the Michell-type structure (see Fig. 4.7d) and the bridge (see Fig. 4.7f). Despite this, the compliance values are similar for the Michell-type structure.

To summarize, the examples in this section demonstrate that using the linear material model ($p = 1$) for the support elements is appropriate for the present simultaneous optimization method, while $p = 3$ may be used to enrich design diversity.

4.3.2 Effect of Mesh Density

To examine the effect of mesh density, three different meshes with element sizes of 0.25 mm, 0.5 mm, and 1 mm are used for the two-bar frame. These are referred to as fine, medium, and coarse meshes, respectively. The results in Fig. 4.8 reveal that different final support locations are obtained from different meshes when the same target number of support elements, $N^* = 6$, is used. The main reason for the variation in support locations is that the six support elements in different meshes correspond to different total support volumes. Specifically, the volume of one element in the coarse mesh is equivalent to the volume of two and four elements in the medium and fine meshes, respectively. Therefore, increasing N^* from 6 to 12 in the fine mesh case leads to identical support locations to those in the medium mesh case. This mesh density study shows that one should be cautious when specifying the number of supports. To avoid mesh dependency of the optimization result, Eq. (4.1c) should be used instead of Eq. (4.2) to impose a constraint on the total volume or cost of support elements.

4.3.3 Effect of Support Stiffness

Support stiffness may have a significant effect on structural compliance and topology. To demonstrate this, we test flexible supports ($E_0 = 0.001 E_1$) and stiff supports ($E_0 = 1000 E_1$) for the Michell-type structure. The results given in Fig. 4.9 show that, compared to the design in Fig. 4.7d generated from $E_0 = E_1$, the compliance is substantially increased with flexible supports and decreased with stiff supports. Distinctly different final structural shapes are obtained. In the flexible support case, the bars connecting the load to the outer arch spread towards end corners to reduce the horizontal movements of the arch near the supports. This topology is similar to the previous evolutionary structural optimization (ESO) solution obtained using a roller support at one corner and a pin at the other, which features two bars aligned with the horizontal line to restrain the horizontal movement of the roller (Xie and Steven 1993). The flexible supports pinned to the base (see Fig. 4.7c, d) act similarly to the pin–roller support conditions. It should be noted that the present example includes additional considerations, such as the size and properties of the supports, in contrast to previous simplified support conditions. On the other hand, the topology generated

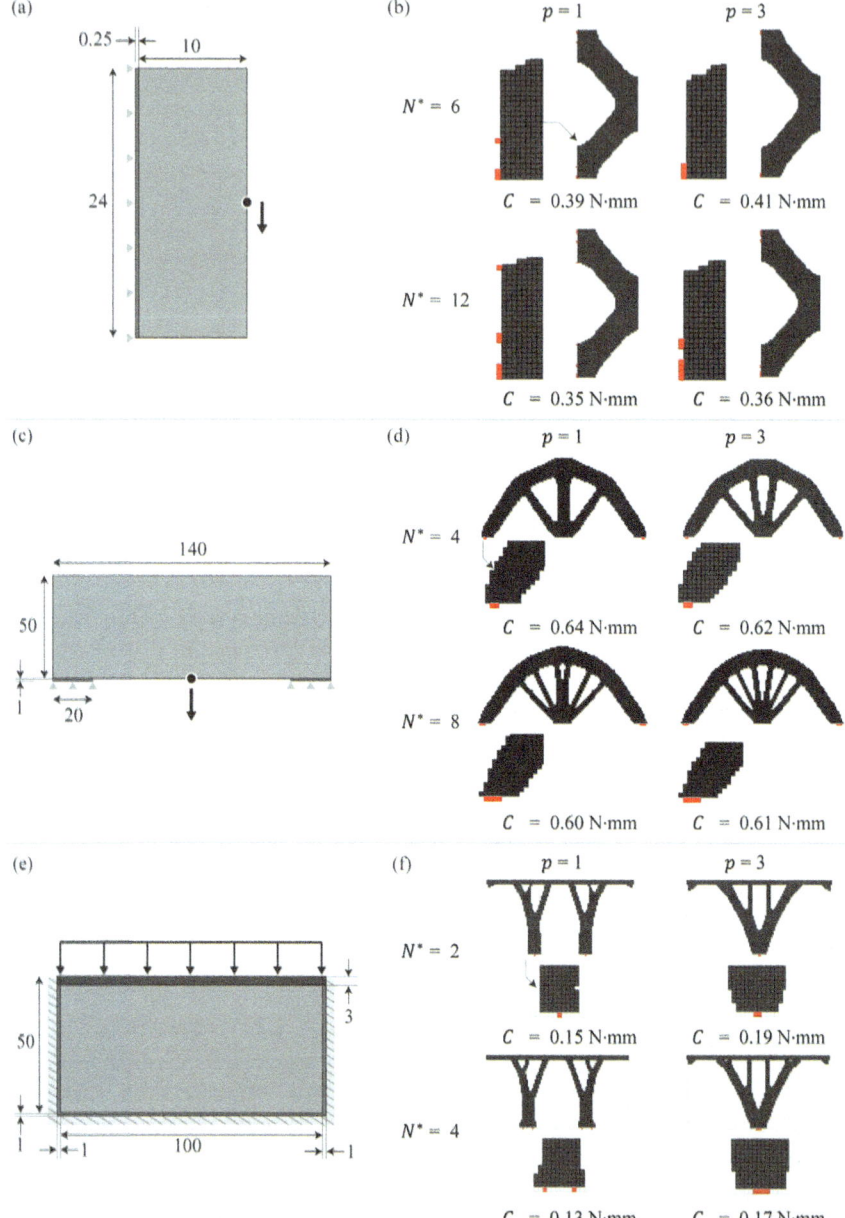

Fig. 4.7 Additional examples to test the effect of using different penalty exponents for the support elements: **a** and **b** two-bar frame; **c** and **d** Michell-type structure; **e** and **f** 2D bridge. Problem definitions are shown in (a), (c) and (e), while optimization results are given in (b), (d), and (f) (reprinted from Lee and Xie 2021, with permission from Elsevier)

4.3 Discussion

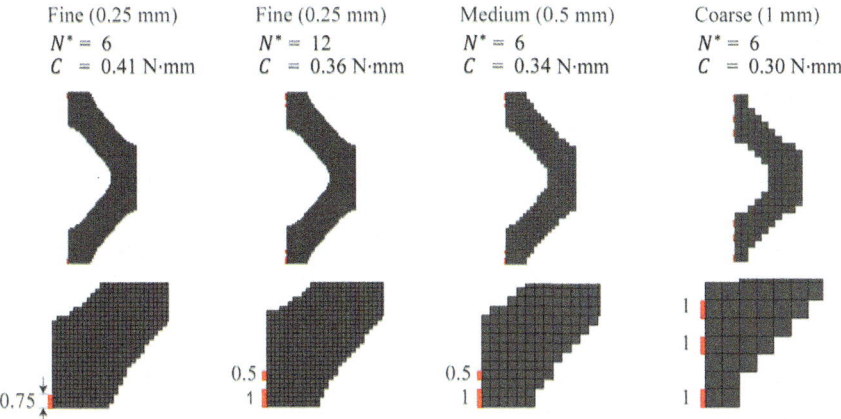

Fig. 4.8 Optimization results of the two-bar frame from using fine, medium, and coarse meshes (reprinted from Lee and Xie 2021, with permission from Elsevier)

with stiff supports has end bars positioned at an angle of 45° to the horizontal line, which bears a strong resemblance to the previous ESO result obtained using rigid pins at the two corners (Xie and Steven 1993).

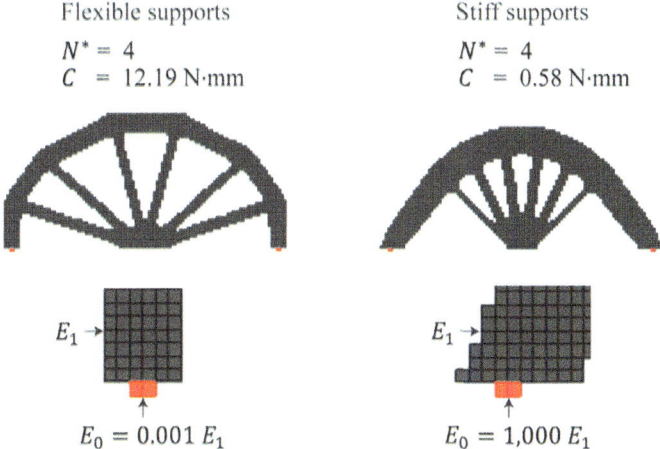

Fig. 4.9 Optimization results of the Michell-type structure for different support stiffness (reprinted from Lee and Xie 2021, with permission from Elsevier)

4.3.4 Effect of Support Cost

Although we have been using Eq. (4.2) to constrain the number of supports, the present simultaneous optimization algorithms can be applied to the more general case of considering the total cost of supports using Eq. (4.1c), with very minor changes to the computer code. The most common measure of the cost is volume. In this case, the cost of each support element, S_i, in Eq. (4.1c) is equal to its volume, V_i.

In some practical applications, the cost of a support may vary significantly from one location to another. Consider the bridge example shown in Fig. 4.7e where the prescribed support domain includes a layer of elements at the bottom and along the two sides. If the cost of every support element is equal and the number of supports is small, the final supports appear at the bottom (see Fig. 4.7f). However, if the cost of supports at the bottom is much higher than that of supports on the sides (as is the case when constructing a bridge over a deep valley), the optimization process will automatically relocate some or all final supports from the bottom to the two sides. Buhl (2002) provided several interesting examples that considered varying costs at different support locations.

4.3.5 Alternative Optimization Methods

We have shown that by adding a layer of support domain to the structural design domain, the optimization of support locations and structural topology can be conducted simultaneously, although the design variables in the two domains are updated using different methods. In this chapter, we have used a standard optimality criteria method to update the design variables for the support domain and the BESO method for the structural design domain.

In fact, other topology optimization methods can also be employed for the structural design domain. For instance, Lee and Xie (2021) introduced a slightly modified SIMP method to replace the BESO method for the structural design domain, while retaining the optimality criteria method for the support domain. Figure 4.10 provides a comparison of optimization results generated from using SIMP and BESO methods for the structural domain. Four structures, including the short cantilever, two-bar frame, Michel-type structure, and 2D bridge, are re-examined here.

In the SIMP method, the design variable x_i can vary continuously between 0 and 1, while in the BESO method, x_i only changes discretely from 1 to x_{min} or vice versa. As shown in Fig. 4.10, the structural topologies produced by the SIMP method exhibit grey elements, particularly during the early stages of the optimization process, while the topologies from the BESO method are always black and white.

It is observed from Fig. 4.10 that using SIMP and BESO methods may result in different final support locations and different topologies, but the values of the objective function (compliance) are the same or very similar. As mentioned earlier, the formation of support locations strongly depends on the structural topology in the

4.3 Discussion

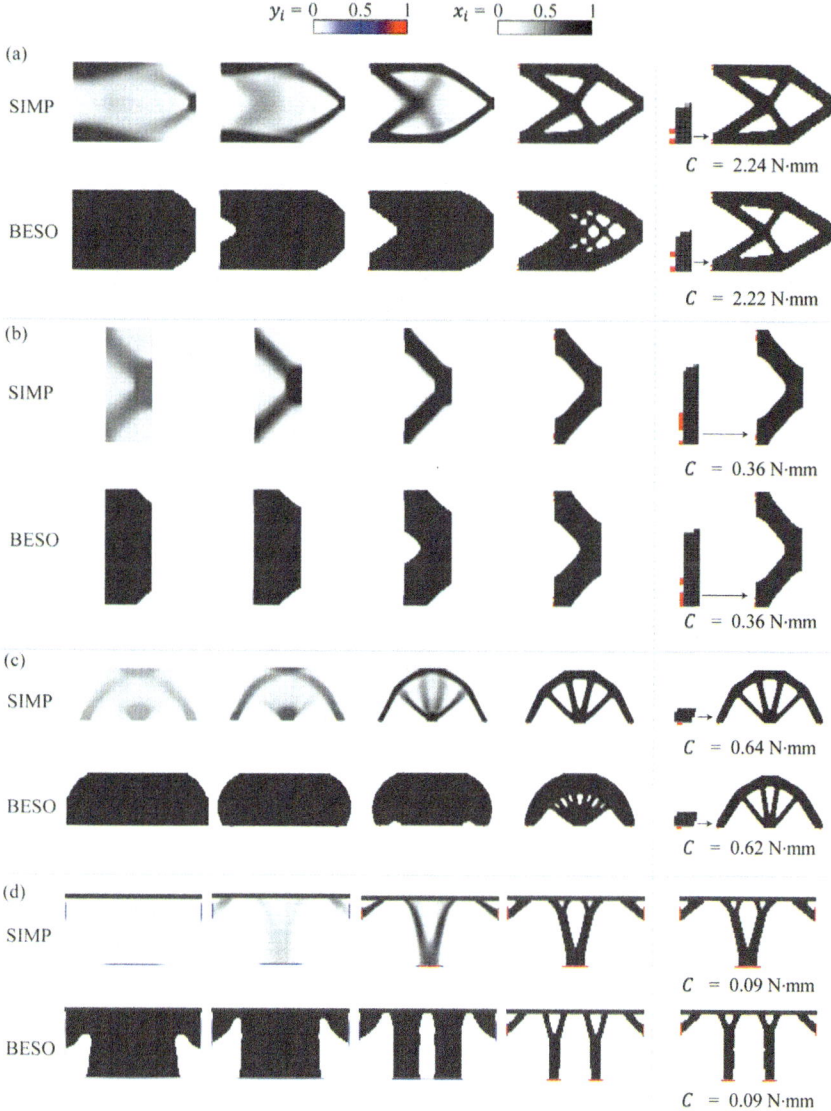

Fig. 4.10 Optimization results from using SIMP and BESO methods for the structural design domain: **a** short cantilever; **b** two-bar frame; **c** Michell-type structure; **d** 2D bridge (reprinted from Lee and Xie 2021, with permission from Elsevier)

early stages of the optimization process. Due to different SIMP and BESO topologies in early iterations, different support locations and structural topologies are ultimately obtained.

Further, in some cases, it is observed that the SIMP method generates slightly higher compliance values than the BESO method. However, this does not necessarily mean that the SIMP method is less suitable for the simultaneous optimization approach. The higher compliance in the SIMP results may have been caused by overestimating the strain energy of the remaining grey elements in the final topologies. It is fair to say that both SIMP and BESO methods can be used effectively for the structural design domain in the simultaneous optimization of support locations and structural topology.

One might ask: can the ESO or BESO method be used for the *support domain*? The answer is no. The simple reason is that the ESO and BESO methods update the design variables discretely, switching from 1 to 0 (or x_{min}), or vice versa. For support optimization problems, the final number of supports is usually small, for example, fewer than 10. Therefore, in later iterations of the optimization process, adding or deleting one support element would make more than 10% change to the remaining support elements. Such a large change contradicts the basic assumption of the ESO and BESO methods, which is that the change at each iteration should be 'evolutionary', typically less than 3% (Huang and Xie 2010).

4.4 Applications

4.4.1 Optimizing Support Locations of a 3D Shell Structure

This example demonstrates a novel application of the support optimization method. Here, the task is to optimize the locations of $N^* = 8$ pin supports for a 3D free-form shell structure. The structure is subjected to a vertical load uniformly distributed along the top edge, as shown in Fig. 4.11a. The bounding dimensions of the structure are approximately 6.9 m × 5.8 m × 3.2 m in length, width, and height, respectively. The thickness of the shell is 10 mm. In this example, the structure itself is treated as non-design domain, and 100 candidate support elements are added to the base of the structure (see Fig. 4.11a). Therefore, the structural topology optimization algorithm is not triggered in this case, and the final support locations are determined using the support optimization algorithm.

Figure 4.11b shows the optimized support locations of the shell structure. To verify whether the optimized solution is indeed superior, we manually shift these supports to random locations (see Fig. 4.11c). A comparison of the compliance values of the optimized solution and the three manual designs is given in Fig. 4.11d. The optimization result has the lowest compliance, indicating that the eight supports are well located for structural efficiency. Shifting these supports to other locations can significantly affect the structural performance. For instance, the compliance of

4.4 Applications

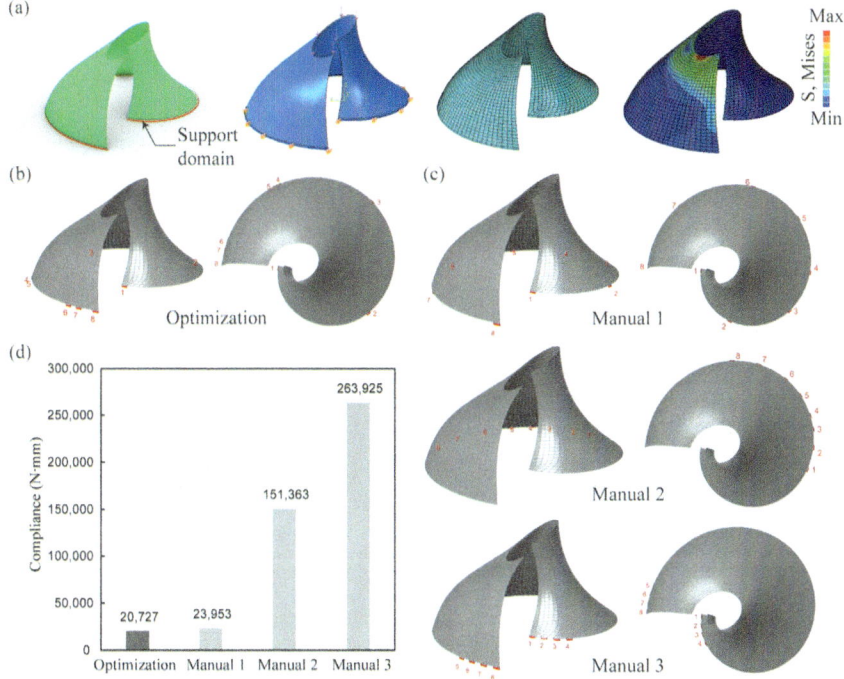

Fig. 4.11 Optimizing support locations of a 3D shell structure: **a** initial setup of the finite element model; **b** optimized support locations; **c** three manual arrangements of support locations; **d** compliance comparison (reprinted from Lee and Xie 2021, with permission from Elsevier)

manual design 3 in Fig. 4.11c is nearly 12 times greater than that of the optimized solution.

This example clearly demonstrates that the performance of a structure can be substantially improved by adjusting its support locations, and the support optimization method can be used to quickly find appropriate support locations.

4.4.2 Designing a 3D Hinge Frame

A 3D hinge frame similar to the one discussed in Sect. 3.6 is reconsidered here. The optimization model, consisting of the boundary and load conditions as well as the design space, is given in Fig. 4.12a, b. Unlike previous studies (Bi et al. 2020; Xiong et al. 2021), the task here is to optimize the locations of $N^* = 4$ support elements in the roller support area (see Fig. 4.12b) while simultaneously optimizing the structural topology.

The optimized design is shown in Fig. 4.12c and its smoothed version is given in Fig. 4.12d. Although the overall structural topology is similar to the previous

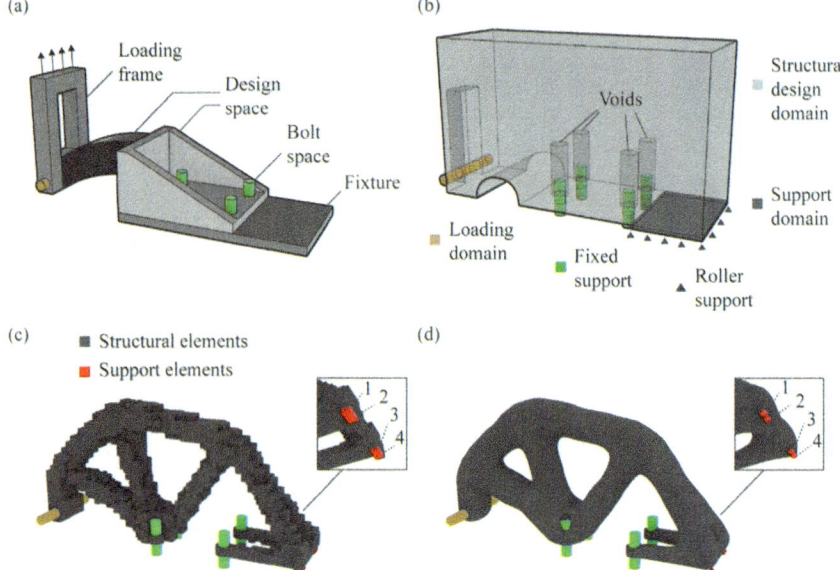

Fig. 4.12 Designing a 3D hinge frame: **a** problem definition; **b** setup of structural design domain and support domain; **c** optimized design; **d** smoothed optimized design (reprinted from Lee and Xie 2021, with permission from Elsevier)

result (see Fig. 3.11), a substantial difference is observed in the front part near the four support elements. This difference occurs because in the previous case the entire bottom surface of the front part can be supported by the base plate through contact, while in this case the front part can only be propped up by four support elements. This example confirms that changes in support conditions can significantly affect the structural shape and that the present simultaneous optimization method can be applied in designing complex 3D structures.

4.5 Conclusion

This chapter has provided a detailed discussion of our method for the simultaneous optimization of support locations and structural topology. Various 2D and 3D examples have been presented to demonstrate the validity, effectiveness, and applicability of this method. The results show that optimizing support locations can substantially improve the structural performance, and that the number of supports and the stiffness of the support material can significantly affect the final topology of the structure.

References

Allaire, G., Jouve, F. and Toader, A. M. (2004) Structural optimization using sensitivity analysis and a level-set method. *J. Comput. Phys.* **194**, 363–393.

Bendsøe, M. P. (1995) *Optimization of Structural Topology, Shape and Material*. Berlin: Springer.

Bi, M., Tran, P. and Xie, Y. M. (2020) Topology optimization of 3D continuum structures under geometric self-supporting constraint. *Addit. Manuf.* **36**, 101422.

Buhl, T. (2002) Simultaneous topology optimization of structure and supports. *Struct. Multidisc. Optim.* **23**, 336–346.

Hu, J., Yao, S. and Huang, X. (2020) Topology optimization of dynamic acoustic–mechanical structures using the ersatz material model. *Comput. Methods Appl. Mech. Eng.* **372**, 113387.

Huang, X. (2021) On smooth or 0/1 designs of the fixed-mesh element-based topology optimization. *Adv. Eng. Softw.* **151**, 102942.

Huang, X. and Xie, Y. M. (2007) Convergent and mesh-independent solutions for the bi-directional evolutionary structural optimization method. *Finite Elem. Anal. Des.* **43**, 1039–1049.

Huang, X. and Xie, Y. M. (2010) *Evolutionary Topology Optimization of Continuum Structures: Methods and Applications*. Chichester: John Wiley & Sons.

Jang, G. W., Shim, H. S. and Kim, Y. Y. (2009) Optimization of support locations of beam and plate structures under self-weight by using a sprung structure model. *J. Mech. Des.* **131**, 021005.

Jihong, Z. and Weihong, Z. (2006) Maximization of structural natural frequency with optimal support layout. *Struct. Multidisc. Optim.* **31**, 462–469.

Kumar, T. and Suresh, K. (2021) Direct Lagrange multiplier updates in topology optimization revisited. *Struct. Multidisc. Optim.* **63**, 1563–1578.

Lee, T.-U. and Xie, Y. M. (2021) Simultaneously optimizing supports and topology in structural design. *Finite Elem. Anal. Des.* **197**, 103633.

Meng, X., Lee, T.-U., Xiong, Y., Huang, X. and Xie, Y. M. (2021) Optimizing support locations in the roof–column structural system. *Appl. Sci.* **11**, 2775.

Rozvany, G. I. N., Zhou, M. and Birker, T. (1992) Generalized shape optimization without homogenization. *Struct. Optim.* **4**, 250–252.

Sigmund, O. (2001) A 99 line topology optimization code written in Matlab. *Struct. Multidisc. Optim.* **21**, 120–127.

Xie, Y. M. and Steven, G. P. (1993) A simple evolutionary procedure for structural optimization. *Comput. Struct.* **49**, 885–896.

Xiong, Y., Bao, D., Yan, X., Xu, T. and Xie, Y. M. (2021) Lessons learnt from a national competition on structural optimization and additive manufacturing. *Curr. Chin. Sci.* **1**, 151–159.

Zelickman, Y. and Amir, O. (2022a) Optimization of column layouts in buildings considering structural and architectural constraints. *Engineering Archive*, https://doi.org/10.31224/2723. Accessed 8 December 2024.

Zelickman, Y. and Amir, O. (2022b) Optimization of plate supports using a feature mapping approach with techniques to avoid local minima. *Struct. Multidisc. Optim.* **65**, 31.

Zuo, Z. H. and Xie, Y.M. (2015) A simple and compact Python code for complex 3D topology optimization. *Adv. Eng. Softw.* **85**, 1–11.

Open Access This chapter is licensed under the terms of the Creative Commons Attribution 4.0 International License (http://creativecommons.org/licenses/by/4.0/), which permits use, sharing, adaptation, distribution and reproduction in any medium or format, as long as you give appropriate credit to the original author(s) and the source, provide a link to the Creative Commons license and indicate if changes were made.

The images or other third party material in this chapter are included in the chapter's Creative Commons license, unless indicated otherwise in a credit line to the material. If material is not included in the chapter's Creative Commons license and your intended use is not permitted by statutory regulation or exceeds the permitted use, you will need to obtain permission directly from the copyright holder.

Chapter 5
Optimizing Load Distributions

Structural optimization is typically performed with predetermined load locations, directions, and magnitudes. This chapter presents ideas and techniques we have developed in recent years to optimize structural designs by treating load distributions as design variables. We discuss methods for optimizing load locations and directions, as well as redistributing load magnitudes. Using this generalized optimization framework, we demonstrate that structural performance (e.g., stiffness) can be significantly enhanced by rearranging the loads. Further, we show that innovative and efficient designs can be achieved through simultaneous optimization of load distribution and structural topology.

5.1 Introduction

Many structures, such as bridges and buildings, are designed to carry loads. Traditionally, structural design optimization is conducted under prescribed load conditions. However, structural performance is often sensitive to the applied loads. Even a minor change in the load distributions—including the load locations, directions, and magnitudes—can significantly influence the structural response. It is highly desirable to apply the loads at the 'sweet spots'—the optimal locations—so that the structure can perform at its best (Cross 1998). Conversely, identifying the most dangerous load distribution is crucial for ensuring the safety of certain structures, such as sports stadiums, which may occasionally experience excessive and unpredictable crowd movements.

The work presented in this chapter challenges the common practice in traditional structural design optimization where load conditions are typically predetermined and fixed. Our approach is particularly relevant and useful for structural design scenarios where load distribution is not fixed, such as managing traffic flow on a multi-lane bridge or stacking objects on a multi-level shelf.

There has been extensive research on the so-called robust optimization that deals with probabilistic (uncertain) load conditions (Takezawa et al. 2011; Jeong et al. 2019). Although probabilistic optimization considers a wide range of load conditions, it can lead to over-designed structures with higher material costs. In contrast, using deterministic loads in structural optimization offers a more straightforward and cost-effective approach, as it is tailored to clearly defined load conditions, particularly the most favourable and detrimental ones.

This chapter focuses on the deterministic optimization of load conditions in structural design. First, we present methods for optimizing load locations. Then, we introduce methods for redistributing load magnitudes and optimizing load directions. These methods can be integrated with topology optimization techniques. Through a series of 2D and 3D examples, we demonstrate the effectiveness of our methods and the benefits of using the generalized optimization approach. Much of the material presented in this chapter is based on the work of Lee and Xie (2022, 2023) and Lee et al. (2025).

5.2 Optimizing Load Locations

5.2.1 Problem Definition and Statement

Suppose we want to distribute a given load vector $\mathbf{F}^*$ equally to a certain number of locations among a larger number of possible candidate load locations. The mathematical problem of compliance minimization (or stiffness maximization) through optimizing load locations in a structure can be formulated as follows

$$\text{Minimize: } C = \mathbf{f}^T \mathbf{u} \tag{5.1a}$$

$$\text{Subject to: } \mathbf{K}\mathbf{u} = \mathbf{f} \tag{5.1b}$$

$$: \mathbf{F}_i = w_i \frac{\mathbf{F}^*}{N^*}, \; i = 1, 2, \ldots, N \tag{5.1c}$$

$$: \sum_{i=1}^{N} w_i = N^*, \; 0 \leq w_i \leq 1 \tag{5.1d}$$

where C is the compliance, $\mathbf{f}$ and $\mathbf{u}$ are the nodal force and displacement vectors, respectively, $\mathbf{K}$ is the global stiffness matrix, N is the total number of candidate load locations, $\mathbf{F}_i$ is the load vector at the ith candidate load location, N^* is the prescribed number of loads in the final design, and w_i is the design variable that denotes the weight of the load at the ith candidate location. The design variable w_i ranges from 0 to 1. At the end of the optimization process, w_i will converge to either 0 or 1 and

5.2 Optimizing Load Locations

$\mathbf{F}_i$ will become 0 or $\mathbf{F}^*/N^*$. In other words, from the N candidate load locations, we aim to find the 'best' N^* locations to equally distribute the given load $\mathbf{F}^*$ so that the structural compliance is minimized.

5.2.2 Sensitivity Analysis

From Eq. (5.1a), the sensitivity of the optimization objective, C, with respect to the design variable, w_i, is

$$\frac{\partial C}{\partial w_i} = \frac{\partial \mathbf{f}^T}{\partial w_i}\mathbf{u} + \mathbf{f}^T\frac{\partial \mathbf{u}}{\partial w_i} \tag{5.2}$$

By taking the derivative of Eq. (5.1b), we get

$$\frac{\partial \mathbf{K}}{\partial w_i}\mathbf{u} + \mathbf{K}\frac{\partial \mathbf{u}}{\partial w_i} = \frac{\partial \mathbf{f}}{\partial w_i} \tag{5.3}$$

During the optimization of load conditions, $\mathbf{K}$ does not change. Therefore, Eq. (5.3) becomes

$$\mathbf{K}\frac{\partial \mathbf{u}}{\partial w_i} = \frac{\partial \mathbf{f}}{\partial w_i} \tag{5.4}$$

$$\mathbf{f}^T\frac{\partial \mathbf{u}}{\partial w_i} = \mathbf{f}^T\mathbf{K}^{-1}\frac{\partial \mathbf{f}}{\partial w_i} \tag{5.5}$$

$$\mathbf{f}^T\frac{\partial \mathbf{u}}{\partial w_i} = \frac{\partial \mathbf{f}^T}{\partial w_i}\mathbf{u} \tag{5.6}$$

Substituting Eqs. (5.6) and (5.1c) into Eq. (5.2) gives

$$\frac{\partial C}{\partial w_i} = 2\frac{\partial \mathbf{f}^T}{\partial w_i}\mathbf{u} = 2\sum_{i=1}^{N}\frac{\partial \mathbf{F}_i^T}{\partial w_i}\mathbf{u}_i = 2\sum_{i=1}^{N}\frac{(\mathbf{F}^*)^T}{N^*}\mathbf{u}_i \tag{5.7}$$

where $\mathbf{u}_i$ is the displacement vector at the ith candidate load location.

5.2.3 Optimization Method

The initial values of the design variables, w_i^0, are set equal to the ratio of N^* to N

$$w_i^0 = \frac{N^*}{N} \tag{5.8}$$

meaning that the total load is equally distributed to all candidate load locations at the start of the optimization process. Then, in each iteration of the optimization process, the continuous design variables, w_i, are updated simultaneously, for $i = 1, 2, \ldots, N$, using an optimality criteria (OC) method similar to the one given by Sigmund (2001), as follows

$$w_i^{\text{new}} = \begin{cases} \max\{0, w_i(1-m)\}, & \text{if } w_i^p B_i^\eta \leq \max\{0, w_i(1-m)\} \\ w_i^p B_i^\eta, & \text{if } \max\{0, w_i(1-m)\} < w_i^p B_i^\eta < \min\{1, w_i(1+m)\} \\ \min\{1, w_i(1+m)\}, & \text{if } w_i^p B_i^\eta \geq \min\{1, w_i(1+m)\} \end{cases} \tag{5.9}$$

where m is the move limit, typically set to 0.2; p is a positive penalty exponent; and η is a coefficient set to -1 here to increase the load magnitudes at candidate load locations with small B_i, which is defined as

$$B_i = \lambda \frac{\partial C}{\partial w_i} \tag{5.10}$$

where λ is a Lagrange multiplier that can be determined using the bisection algorithm discussed in Sect. 4.2.2. It is noted that using $p > 1$ can amplify the differences between design variables without changing their relative ranking and, more importantly, improve the convergence by pushing w_i towards 0 or 1. The iterative optimization process continues until the convergence criterion given in Eq. (4.17) is satisfied.

5.2.4 Optimization Procedure

The iterative procedure for optimizing load locations is as follows:

1. Initialization: Discretize the design space using a finite element mesh, assign candidate load locations at selected nodes, and equally distribute the total load $\mathbf{F}^*$ to all candidate locations using Eqs. (5.1c) and (5.8).
2. Structural analysis: Perform finite element analysis (FEA) to generate information about each candidate load location, such as the displacement vector $\mathbf{u}_i$.
3. Sensitivity analysis: Calculate the sensitivity value of each candidate load location using Eq. (5.7).
4. Optimization: Update each design variable w_i using Eq. (5.9).
5. Convergence check: Repeat Steps 2–4 until the convergence criterion given in Eq. (4.17) is satisfied.

Note that in Step 4, multiple optimizers can operate simultaneously, including the optimization of load locations, load directions, support locations, and structural

5.2 Optimizing Load Locations

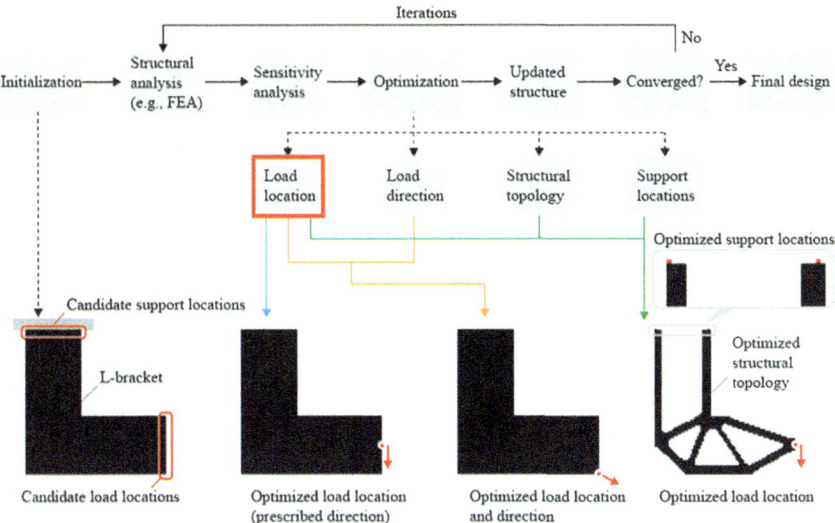

Fig. 5.1 Illustration of the computational workflow of optimizing load locations, which can be combined with optimization of load directions, structural topology, and support locations (reprinted from Lee and Xie 2022, with permission from Elsevier)

topology, as illustrated in Fig. 5.1. The L-bracket is supported along the top edge and loaded on the right edge. If the loads are restricted to the vertical direction and the target number of loads N^* is set to 1, we find that the optimized load location for minimizing the compliance is at the midpoint of the right edge. If the vertical load is moved to the top or bottom corner, the compliance would increase by 6.94% and 6.93%, respectively. Next, we find that optimizing both load location and direction leads to a different load location (at the bottom right corner), which reduces the compliance by 87.10% compared with the optimization solution for the vertically loaded case. Further, if we perform simultaneous optimization of the load location, structural topology, and support locations, we obtain the result shown at the bottom right of Fig. 5.1. More examples of simultaneous optimization of load conditions and structural topology are given in Sects. 5.2.5 and 5.4.4, while details about optimizing load directions are provided in Sect. 5.5.

5.2.5 Numerical Examples

A 2D tree-like structure shown in Fig. 5.2 is 1,823.57 mm tall and 1,534.73 mm wide, with a vertical trunk 128.18 mm wide. The structure is discretized using triangular elements with a mesh size of 3 mm and a thickness of 1 mm. The material is assumed to be linearly elastic and isotropic, with Young's modulus of 100 GPa and Poisson's ratio of 0.30. A fixed boundary condition is applied at the base of the trunk. All

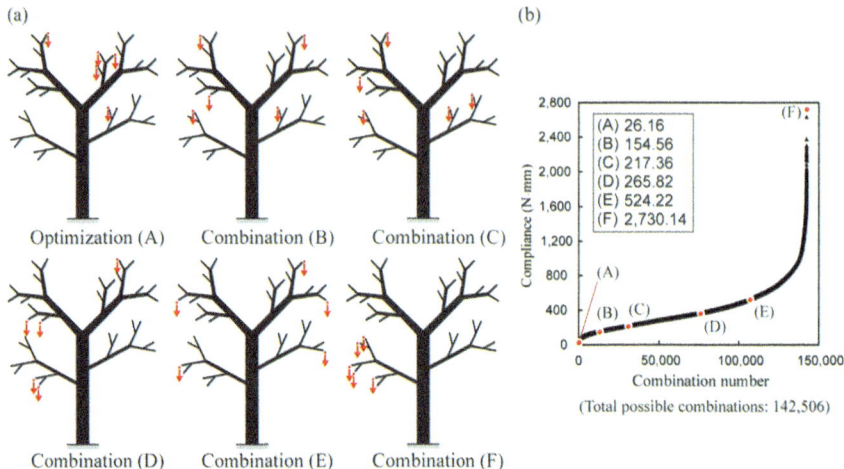

Fig. 5.2 Finding the combination of five load locations that minimizes the compliance of the tree-like structure: **a** the optimized load locations (A) and five manually selected combinations (B)–(F); **b** a comparison of all 142,506 possible combinations, showing that the optimization result has the lowest compliance (reprinted from Lee and Xie 2022, with permission from Elsevier)

loads act in the vertical downward direction, with a total magnitude $F^* = |\mathbf{F}^*| = 100$ N. Thirty candidate load locations are assigned at the tips of all branches. Other parameters for the load location optimization are $N^* = 5$, $\eta = -1$, $m = 0.2$, and $p = 2$.

We use the optimizer to find a combination of five load locations at branch tips that minimizes structural compliance. Note that each of the five loads has the same magnitude (20 N). Searching for the best combination of five load locations among 30 branch tips is not an easy task, as there is a total of $30!/(5!25!) = 142{,}506$ possible combinations. However, our optimizer successfully finds the optimal load locations in fewer than 100 iterations and within 30 min on an ordinary personal computer, despite the use of a very fine mesh for the FEA model. To verify the solution from the optimizer, we have tested all possible combinations of five loads (see Fig. 5.2b) using an efficient, exhaustive search method (discussed in Sect. 5.3). It is found that the optimization result (see (A) in Fig. 5.2a) indeed has the lowest compliance. This example demonstrates the effectiveness of our method for optimizing load locations to enhance structural performance.

In the next example, shown in Fig. 5.3, we aim to design an efficient and innovative 3D structure to carry a uniformly distributed load (UDL) on the top surface while resisting two additional point loads on the front face. The structure is 262.50 mm high, 202.50 mm deep, and 80 mm wide (see Fig. 5.3a). Due to symmetry, only half of the structure needs to be modelled. The half structure is discretized using solid cubic elements that have an edge length of 2.50 mm. The material is the same as that used in the previous tree-like structure. A fixed boundary condition is applied to the back of the entire vertical plate. The horizontal plate at the top is subjected to

5.2 Optimizing Load Locations

a vertical UDL with a total magnitude of 5 kN for the half model. A total of 1,296 candidate load locations are assigned on the front face of the half design domain. Load location optimization parameters for the half model are $N^* = 1$, $\eta = -1$, $m = 0.2$, and $p = 2$. The BESO topology optimization parameters are $ER = 3\%$ and $r_{min} = 8$ mm. The volume fraction is set to 10%. Throughout the optimization process, the non-design domain, which includes the vertical plate at the back and the horizontal plate at the top, remains unchanged. The final structural topologies are smoothed using the Laplacian surface editing technique (Sorkine et al. 2004).

Figure 5.3b shows three optimized designs corresponding to different magnitudes of the vertical point load: $F^* = 2 \times 1$ kN, 2×5 kN, and 2×15 kN. We find that the magnitude of the point load relative to the value of the UDL significantly affects the optimized load locations and the resulting structural topology. When F^* is larger, the optimized load locations are closer to the horizontal plate, and the structural topology is dominated by the point loads, leading to a final shape similar to the classic two-bar frame (Xie and Steven 1997). Conversely, when F^* is smaller, the structural topology is dominated by the UDL, resulting in many branches that support the horizontal plate.

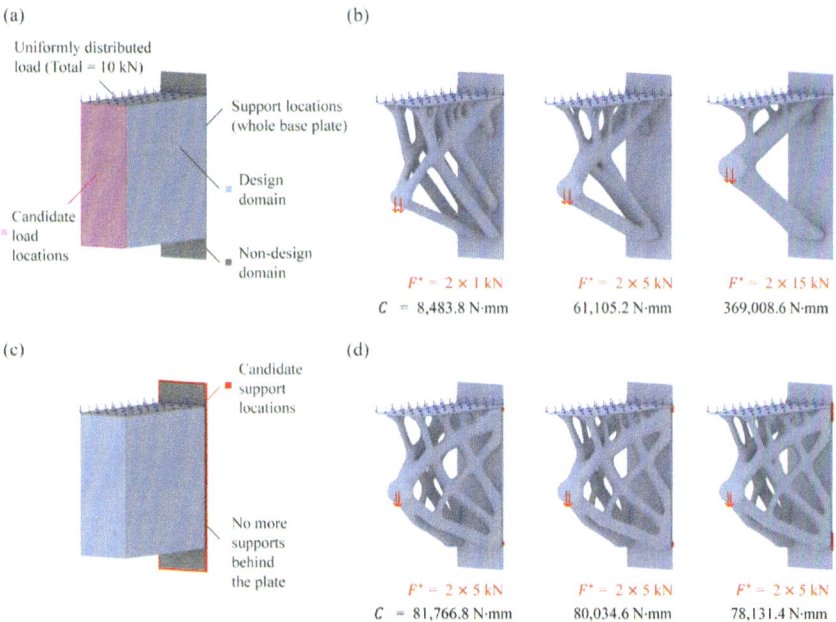

Fig. 5.3 Designing a 3D structure through simultaneous optimization: **a** initial setup for simultaneously optimizing the point load locations and the structural topology; **b** optimization results for three different point load magnitudes; **c** new boundary conditions for including support locations in the optimization; **d** optimization results for three different final support areas (reprinted from Lee and Xie 2022, with permission from Elsevier)

The results in Fig. 5.3b are obtained under the assumption that the entire base plate is fixed to the wall. If the support conditions are allowed to vary, we can employ the simultaneous optimization technique to optimize load locations, support locations, and structural topology, thereby providing unprecedented design freedom in structural optimization. To this end, we remove the original supports behind the vertical plate and include 139 candidate support locations (in the half model) around the plate as additional design variables in the new optimization model, as shown in Fig. 5.3c. This allows us to optimize support locations by removing inefficient supports and preserving efficient ones, using the method presented in Chap. 4. Figure 5.3d shows three new designs with the number of support elements in the half model set to 4, 7, and 21, respectively. It is evident that the optimized three topologies share overall similarities but differ in details. Prescribing larger support areas results in stiffer structures, corresponding to lower compliance. The new structures in Fig. 5.3d are quite different from the designs in Fig. 5.3b in both topology and compliance, demonstrating that support conditions can substantially influence the resulting structural topology and lead to significant variation in the structural performance.

The example in Fig. 5.3 clearly illustrates the benefits and new opportunities of considering load locations, support locations, and structural topology as design variables simultaneously in the optimization process.

5.3 Finding Globally Optimal Locations of Multiple Point Loads Using a Single FEA

While developing the iterative optimization algorithm for optimizing load locations (Lee and Xie 2022), we discovered an exhaustive search method that can find globally optimal locations of multiple point loads at a low computational cost. In fact, it can find the global optimum after performing a single FEA (Lee and Xie 2023). Although a structural design with slightly worse performance than the absolute best is acceptable for most engineering applications, a globally optimal solution is theoretically important as it can serve as a benchmark for verifying the effectiveness and accuracy of numerical algorithms. Indeed, by using the *single FEA method*, we can quickly obtain the compliance values for all the 142,506 possible load combinations for the tree-like structure shown in Fig. 5.2. Interestingly, the solution obtained from the iterative OC method is identical to the globally optimal design from the single FEA method. This result provides clear evidence of the effectiveness of the optimization method given in Sect. 5.2. However, it does not mean we can guarantee that a gradient-based iterative optimization algorithm will always reach the global optimum. We present details of the single FEA method below and demonstrate its potential applications through two examples.

5.3.1 Condensed Flexibility Matrix

In Eq. (5.1b), the dimensions of the global stiffness matrix $\mathbf{K}$, the displacement vector $\mathbf{u}$, and the applied force vector $\mathbf{f}$ are $L \times L$, $L \times 1$, and $L \times 1$, respectively, where L is the degrees of freedom of the finite element model. Usually, L is a large number. When optimizing load locations, the number of candidate load locations N is typically only a small fraction of L; for example, in the tree-like structure shown in Fig. 5.1, N is less than 0.1% of L. Therefore, most components of the force vector $\mathbf{f}$ are zero, except for those corresponding to the candidate load locations. Instead of dealing with the large global matrices and vectors shown in Eqs. (5.1a to 5.1d), we can calculate the compliance more efficiently by focusing only on the candidate load locations, as follows

$$C = \mathbf{f}^T \mathbf{u} = \mathbf{P}^T \mathbf{U} \tag{5.11}$$

where $\mathbf{P}$ and $\mathbf{U}$ are the force and displacement vectors, respectively, associated with the candidate load locations. The relationship between $\mathbf{P}$ and $\mathbf{U}$ is given below

$$\mathbf{U} = \mathbf{DP} \tag{5.12}$$

where $\mathbf{D}$ is the *condensed flexibility matrix*, which is a subset of the global flexibility matrix $\mathbf{K}^{-1}$. Theoretically, $\mathbf{D}$ can be obtained by removing all the rows and columns in $\mathbf{K}^{-1}$ that are not related to the candidate load locations. The dimensions of $\mathbf{D}$ and $\mathbf{P}$ are $M \times M$ and $M \times 1$, respectively, where

$$M = aN \tag{5.13}$$

and a is the number of allowable load directions at each candidate location. If all the loads must be applied in the same direction, $a = 1$; if the loads are allowed to act in x, y, and z directions, $a = 3$.

Numerically, we can calculate the $\mathbf{D}$ matrix by conducting a single FEA of a structure with multiple load cases. This process is not computationally demanding since the $\mathbf{K}$ matrix only needs to be decomposed once, regardless of the number of load cases. To obtain the $\mathbf{D}$ matrix, a total of M load cases is considered. For each load case, the equilibrium Eq. (5.1b) is solved by applying a unit load (with a magnitude of 1) at one of the candidate load locations in one of the allowed directions, while setting all other components in $\mathbf{f}$ to zero. The resulting displacements at the candidate load locations form one column of the $\mathbf{D}$ matrix, corresponding to the degree of freedom of the applied unit load. The complete $\mathbf{D}$ matrix is obtained once all M load cases have been considered, with each column corresponding to a unique load case.

5.3.2 Possible Load Combinations or Permutations

Once the **D** matrix is obtained from the single FEA, we can quickly calculate the compliance C for every possible load **P** using the following equation

$$C = \mathbf{P}^T \mathbf{D} \mathbf{P} \tag{5.14}$$

As the prescribed number of loads is N^*, there should be only N^* non-zero components in each **P**. If the prescribed loads are of the same magnitude and in the same direction (as in the tree-like structure example shown in Fig. 5.2), the total number of possible load arrangements, P_C, can be calculated using the *combination* equation

$$P_C = \frac{M!}{(M - N^*)! N^*!} \tag{5.15}$$

On the other hand, if the prescribed loads have different magnitudes or directions (as in the example shown in Fig. 5.5), changing the order of the applied loads results in a different structural response. In this case, the total number of possible load arrangements, P_P, should be calculated using the *permutation* equation

$$P_P = \frac{M!}{(M - N^*)!} \tag{5.16}$$

After considering all the load combinations or permutations, we obtain an exhaustive list of compliance values corresponding to every possible load arrangement. From this list, we can immediately identify the globally optimal load locations (for both compliance minimization and maximization). Moreover, the single FEA method can be integrated with topology optimization techniques to enable simultaneous optimization of load locations and structural topology.

5.3.3 Numerical Examples

We demonstrate potential practical applications of the single FEA method using a complex 3D structure shown in Fig. 5.4a. This example represents a pavilion design featuring a doubly curved roof, a flat base, and three hollow columns. Here, we search for the extremal arrangements of five vertical loads that yield the lowest or highest structural compliance (C) when the five loads are applied among 37 candidate locations. This design problem can also be viewed as finding the 'best' and 'worst' standing locations on the curved rooftop when it supports five people of the same weight.

In detail, the bounding dimensions of the 3D structure are approximately 500 mm × 500 mm × 259 mm. The structure is discretized using quadrilateral shell elements

5.3 Finding Globally Optimal Locations of Multiple Point Loads Using …

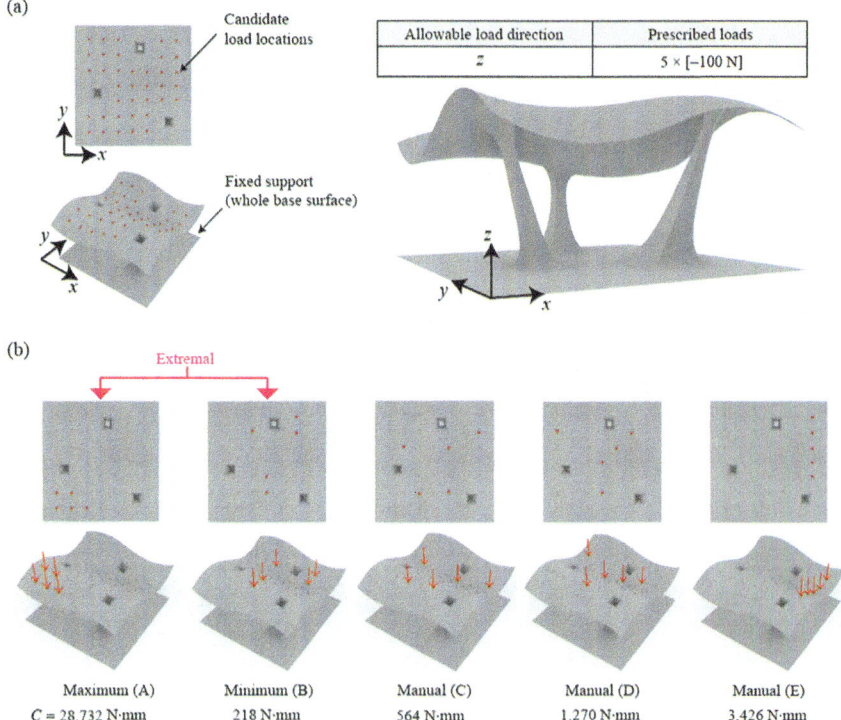

Fig. 5.4 Arrangements of five identical loads among 37 candidate locations in a 3D structure with a doubly curved roof: **a** initial setup of the design problem; **b** five results shown from left to right: highest compliance, lowest compliance, and three manually selected load arrangements (reprinted from Lee and Xie 2023, licensed under CC-BY 4.0)

that have a mesh size of 5 mm and a thickness of 1 mm. The material is assumed to be linearly elastic and isotropic, with Young's modulus of 100 GPa and Poisson's ratio of 0.30. A fixed boundary condition is applied to the whole base surface. Thirty-seven candidate load locations are assigned on the rooftop ($N = 37$). The prescribed loads consist of five vertical forces, each with a magnitude of -100 N ($N^* = 5$). These loads are allowed only in the z direction ($a = 1$). Hence, the number of load cases required in the single FEA is $M = 1 \times 37 = 37$, and the total number of possible arrangements of the prescribed loads is $P_C = 37!/(32!5!) = 435,897$.

Figure 5.4b shows the calculation results for the structure under various load conditions. It is reasonable to observe that the five load locations to achieve the maximum compliance are clustered together near the roof edge, corresponding to the 'worst' standing locations. In contrast, to achieve the minimum compliance, three loads are distributed around the shortest column, while the remaining two are positioned in the 'mountain' region of the rooftop between the two long columns. These correspond to the 'best' five standing locations. Figure 5.4b clearly shows that altering the arrangement of load locations significantly affects structural compliance.

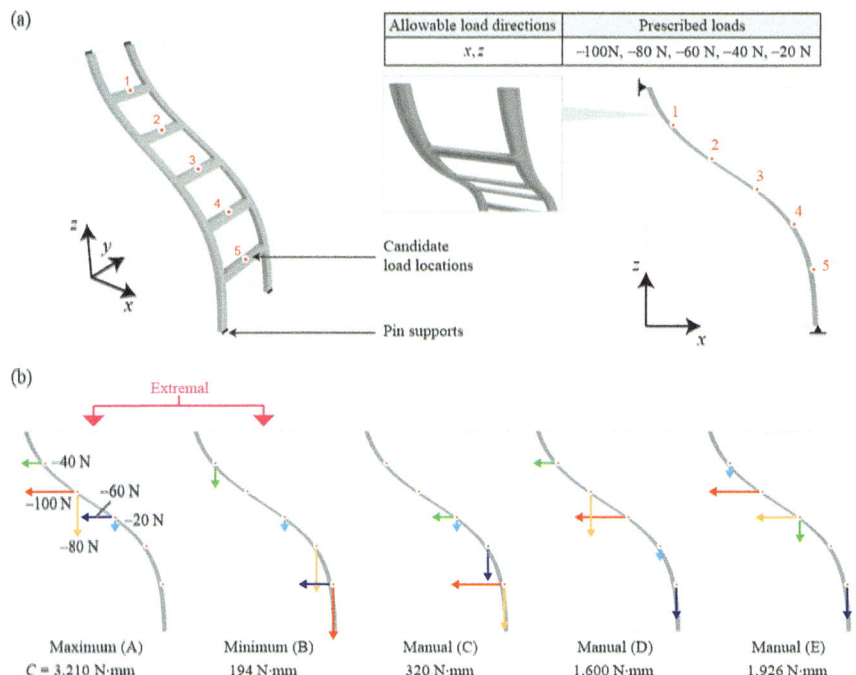

Fig. 5.5 Arrangements of five loads of different magnitudes among five candidate locations in two possible directions: **a** initial setup of the design problem; **b** five results shown from left to right: highest compliance, lowest compliance, and three manually selected load arrangements (reprinted from Lee and Xie 2023, licensed under CC-BY 4.0)

Although there are 435,897 possible load combinations, only a single FEA with 37 load cases is required to determine the globally optimal load locations.

In the previous examples shown in Figs. 5.3 and 5.4, the structures are designed to carry identical loads. However, some structures may be required to carry loads with different magnitudes and signs. The single FEA method can readily handle such cases by following the same procedure as outlined above. This is demonstrated with an example of a 3D curved structure shown in Fig. 5.5, where the prescribed loads have different magnitudes. In this example, we search for the extremal arrangements of five loads among five candidate locations which yield the highest and lowest structural compliance. The bounding dimensions of the structure are approximately 299 mm × 161 mm × 405 mm in the x, y, and z directions. The structure is tapered from the bottom to the top, with members featuring an approximately 15 mm wide and 10 mm deep 'C' profile (see Fig. 5.5a). The structure is discretized using quadrilateral shell elements, with a mesh size of 2 mm and a thickness of 1 mm. The material is assumed to be linearly elastic and isotropic, with Young's modulus of 100 GPa and Poisson's ratio of 0.30. Pinned boundary conditions are applied to four end edges—two at the bottom and two at the top. Five candidate load locations ($N = 5$) are assigned at midpoints of the five horizontal members. The five prescribed loads ($N^* = 5$) are −

100 N, − 80 N, − 60 N, − 40 N, and − 20 N, which are allowed to be applied in x and z directions ($a = 2$). Thus, the number of load cases required in the single FEA is $M = 2 \times 5 = 10$, and the total number of possible arrangements of the prescribed loads is $P_P = 10!/5! = 30{,}240$.

Figure 5.5b presents the calculation results for the structure under various load arrangements. It is noted that some load locations carry two loads in different directions. Despite having five candidate load locations and five prescribed loads, not all candidate locations need to carry a load, since each location can accommodate up to two loads in different directions ($a = 2$). To reach the maximum compliance, the prescribed loads are seen to be gathered on the top part of the structure, while the remaining candidate locations are not carrying any loads. In contrast, to achieve the minimum compliance, most loads are positioned near the two support regions, with the larger forces gathered around the lower, stiffer part of the structure.

It should be noted that although the loads in this example can act in two directions, we are not optimizing the direction of the resultant force at each candidate load location, since the magnitudes of all load components in x and z directions are predetermined. The optimization of load directions is discussed in Sect. 5.5.

5.4 Redistributing Load Magnitudes

In previous discussions, the magnitudes of all loads are assumed either to be equal (Sect. 5.2) or to have fixed values (Sect. 5.3). In this section, we present methods for finding the extremal distribution of load magnitudes to minimize or maximize structural compliance (Lee et al. 2025). Specifically, we introduce a modified OC method tailored for the problem of optimizing the distribution of load magnitudes. Furthermore, we show that this problem can be effectively solved using two well-established mathematical programming methods. The efficacy of these methods is demonstrated through two numerical examples, of which one illustrates the simultaneous optimization of the distribution of load magnitudes and the structural topology.

5.4.1 Modified Optimality Criteria Method

To optimize the distribution of load magnitudes, we modify the formulation from Eqs. (5.1a to 5.1d) as follows

$$\text{Minimize}: \ C = \mathbf{f}^T \mathbf{u} = \mathbf{P}^T \mathbf{DP} \tag{5.17a}$$

$$\text{Subject to}: \ \mathbf{Ku} = \mathbf{f} \tag{5.17b}$$

$$: \mathbf{f}(\mathbf{P}) = \mathbf{Q}^* \tag{5.17c}$$

$$: P_{\min} \leq P_i \leq P_{\max}, \ i = 1, 2, \ldots, M \qquad (5.17d)$$

where $\mathbf{P} = [P_1, P_2, \ldots, P_M]^T$ is the load vector to be optimized, $P_{\min}$ and $P_{\max}$ are the prescribed bounds for each load component P_i, $\mathbf{f}(\mathbf{P})$ represents a function of $\mathbf{P}$, and $\mathbf{Q}^*$ contains prescribed target values for $\mathbf{f}(\mathbf{P})$. If vector $\mathbf{Q}^*$ has a dimension of $G \times 1$, Eq. (5.17c) represents G groups of constraints on the load vector $\mathbf{P}$. When these constraints are linear functions of $\mathbf{P}$, Eq. (5.17c) can be written as

$$\mathbf{WP} = \mathbf{Q}^* \qquad (5.18)$$

where $\mathbf{W}$ is a $G \times M$ matrix of weighting factors (which can be any constants). For example, as a special case, we may require the total forces in x, y, and z directions to be equal to prescribed values, Q_1^*, Q_2^*, and Q_3^*. In this case, the linear constraints in Eq. (5.18) can be further simplified to

$$\sum_{j=1}^{N} P_{j,x} = Q_1^* \qquad (5.19a)$$

$$\sum_{j=1}^{N} P_{j,y} = Q_2^* \qquad (5.19b)$$

$$\sum_{j=1}^{N} P_{j,z} = Q_3^* \qquad (5.19c)$$

where $P_{j,x}$, $P_{j,y}$, and $P_{j,z}$ are the x, y, and z components of the load at candidate location j. Similar to the sensitivity analysis given in Sect. 5.2.2, we can obtain the sensitivity of the objective function with respect to the design variable $\mathbf{P}$ from Eq. (5.17a), as follows

$$\boldsymbol{\alpha} = \frac{\partial C}{\partial \mathbf{P}} = 2\mathbf{DP} \qquad (5.20)$$

Note that for load magnitude optimization, the sensitivity $\boldsymbol{\alpha}$ from Eq. (5.20) often contains both positive and negative values, which is unsuitable for the standard OC method (Sigmund 2001) commonly used for topology optimization. To apply the OC method here, a few modifications are necessary. First, we normalize the components of $\boldsymbol{\alpha}$

$$\bar{\alpha}_i = \frac{\alpha_i}{|\alpha_i|_{\max}} \qquad (5.21)$$

so that $\bar{\alpha}_i$ is in the range of $[-1, 1]$. Second, we change the original design variable P_i to

5.4 Redistributing Load Magnitudes

$$\overline{P}_i = P_i - P_{\min} \tag{5.22}$$

so that the revised design viable $\overline{P}_i$ is always greater than or equal to zero. In each iteration of the optimization process, $\overline{P}_i$ is updated simultaneously for $i = 1, 2, ..., M$ using a modified optimality criteria (MOC) method, as follows

$$\overline{P}_i^{\text{new}} = \begin{cases} \max(0, \overline{P}_i - m_p), & \text{if } \overline{P}_i \overline{B}_i \leq \max(0, \overline{P}_i - m_p) \\ \overline{P}_i \overline{B}_i, & \text{if } \max(0, \overline{P}_i - m_p) < \overline{P}_i \overline{B}_i < \min(P_{\max} - P_{\min}, \overline{P}_i + m_p) \\ \min(P_{\max} - P_{\min}, \overline{P}_i + m_p), & \text{if } \overline{P}_i \overline{B}_i \geq \min(P_{\max} - P_{\min}, \overline{P}_i + m_p) \end{cases} \tag{5.23}$$

where m_p is the move limit (more details later), and $\overline{B}_i$ is a scaling factor that can be calculated using the normalized sensitivity value $\overline{\alpha}_i$ from Eq. (5.21)

$$\overline{B}_i = \frac{\gamma^{-\overline{\alpha}_i}}{\lambda} \tag{5.24}$$

where $\gamma > 1$ is a projection constant, set to 2 in this chapter. Here $\gamma^{-\overline{\alpha}_i}$ is employed to project the sensitivity value to $[\frac{1}{\gamma}, \gamma]$ so that negative values are avoided. The Lagrange multiplier λ can be determined using the bisection algorithm discussed in Sect. 4.2.2. However, instead of the structural volume constraint, here the linear constraint groups in Eq. (5.18) are applied to determine λ.

The move limit m_p can be calculated as

$$m_p = \left| \frac{Q_n^* \times ER_p}{M} \right|, \ n = 1, 2, ..., G \tag{5.25}$$

where ER_p is the evolutionary rate for load optimization, set to 0.5%. It is important to note that if multiple constraint groups are introduced (i.e., $G > 1$), Eqs. (5.23–5.25) need to be performed G times, with the loads in each constraint group updated independently. To obtain a unique λ for each constraint group, load components in one constraint group should not appear in other constraint groups. A typical example of such 'independent' load constraint groups can be found in Eqs. (5.19a to 5.19c).

The iterative optimization process continues until the convergence criterion given in Eq. (4.17) is satisfied.

5.4.2 Sequential Least Squares Quadratic Programming Method

Like most other gradient-based numerical search methods, the MOC method presented in the previous section may not always converge to the global optimum.

Alternatively, mathematical programming methods can be used to solve the optimization problem given in Eqs. (5.17a to 5.17d). One of the well-established mathematical programming methods is the sequential least squares quadratic programming (SLSQP) method (Kraft 1988), which is particularly well suited for handling the quadratic relationship between the objective function C and the design variable **P** in Eq. (5.17a). Moreover, the SLSQP method is highly versatile, capable of tackling both linear and non-linear constraints in Eq. (5.17c). The SLSQP optimizer we have used in this chapter is from the SciPy open-source software package written in Python (SciPy 2024).

5.4.3 Interior-Point Method

If the constraints in Eq. (5.17c) are linear functions of the design variable **P**, as in Eqs. (5.18) and (5.19a to 5.19c), the optimization problem of Eq. (5.17) can be solved effectively and efficiently using the interior-point (INP) method (Nesterov and Nemirovskii 1994). As discussed in Sect. 5.3.1, the condensed flexibility matrix **D** is a subset of the global flexibility matrix $\mathbf{K}^{-1}$. Since $\mathbf{K}^{-1}$ must be positive definite for a stable structure, **D** is also positive definite. Therefore, Eq. (5.17a) represents a convex quadratic objective function of **P**. With the additional condition of linear constraints in Eq. (5.17c), this special optimization setting ensures that the solution from the INP method is the global optimum. The INP method is well suited for this type of convex optimization problems, as it provides higher robustness and efficiency than the SLSQP method (Goulart and Chen 2024). Notably, unlike the MOC and SLSQP methods that require an initial guess for **P**, the INP method can operate without an initial design. The INP optimizer we have used in this chapter is part of the MOSEK software package (MOSEK 2019).

5.4.4 Numerical Examples

To demonstrate potential practical applications of optimizing the distribution of load magnitudes, we first consider a multi-level shelf structure shown in Fig. 5.6. Suppose we need to store a certain amount of material on the shelf and the material can be divided into small bags and placed across $N = 27$ candidate locations (see Fig. 5.6a) to minimize the compliance, C. If the total weight of the material is 200 N, the constraint on the loads given in Eq. (5.17c) takes its linear form as shown in Eq. (5.19c), i.e., $\sum_{j=1}^{N} P_{j,z} = -200$.

The shelf is designed to be 80 mm wide, 40 mm deep, and 140 mm high, and it is discretized using triangular shell elements with a mesh size of 1.5 mm and a thickness of 1 mm. The material is assumed to be linearly elastic and isotropic,

5.4 Redistributing Load Magnitudes

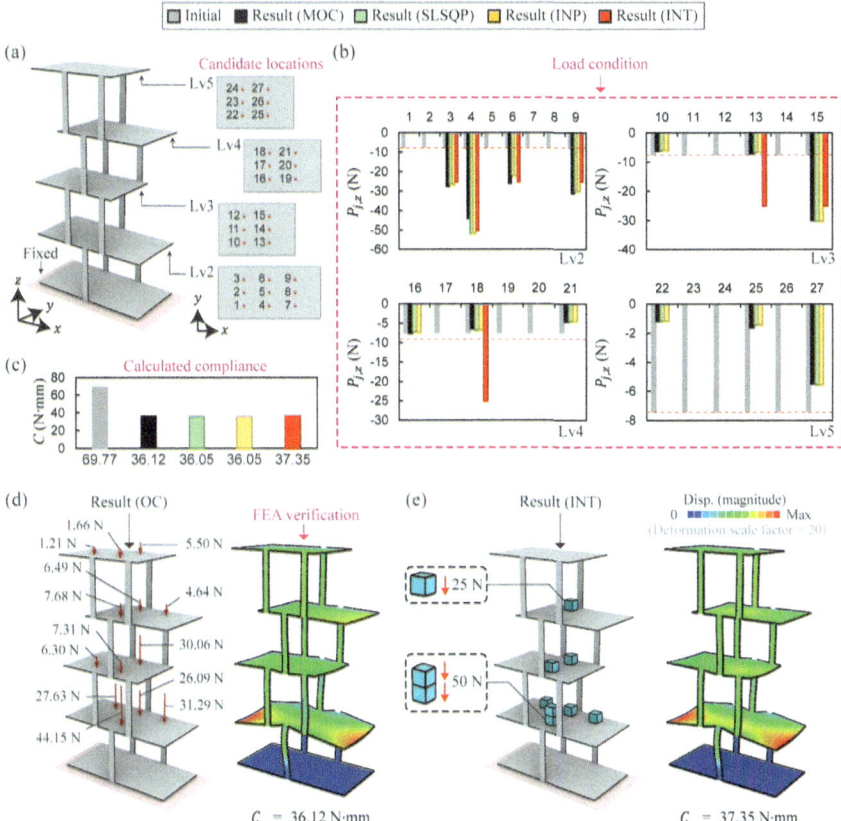

Fig. 5.6 Distributing a total weight of 200 N across 27 candidate locations on a multi-level shelf structure to minimize its compliance, C: **a** candidate load locations on four levels of the shelf; **b** initial and optimized distributions of load magnitudes obtained from different methods; **c** comparison of the C values; **d** FEA verification of the optimization result from the OC method; **e** optimizing the placement of eight boxes, each weighing 25 N (reprinted from Lee et al. 2025, licensed under CC-BY 4.0)

with Young's modulus of 100 GPa and Poisson's ratio of 0.30. The shelf has five levels, with nine, six, six, and six candidate load locations assigned to Levels 2–5, respectively (see Fig. 5.6a). A fixed boundary condition is applied to the base. The loads are constrained between $P_{\min} = -200$ N and $P_{\max} = 0$, allowing only downward forces. In the initial guess design, we distribute the total load of 200 N equally to the 27 candidate locations, which results in $C = 69.77$ N·mm.

Figure 5.6b–d illustrates the optimization results. The magnitudes of individual loads at the 27 locations are shown in Fig. 5.6b, revealing that the optimized distributions of load magnitudes from the MOC, SLSQP, and INP methods are very similar. A comparison of the compliance values corresponding to the initial and optimized designs is given in Fig. 5.6c. It is observed that by optimizing the distribution of load magnitudes, the compliance decreases by almost 50%, which represents a huge improvement in the overall stiffness of the structure that carries the same amount of total weight. Due to the convex nature of the optimization problem and the linear constraint in this case, the result from the INP method must be the global optimum, with $C = 36.05$ N·mm. Although the result from the SLSQP cannot be guaranteed to be the global optimum, in this particular case the obtained C value also reaches 36.05 N·mm. Further, it is noted that the OC method performs quite well, achieving an optimized C value of 36.12 N·mm, only 0.19% higher than the global optimum.

Now, let us consider a slightly different scenario where the material is packed in eight boxes, each weighing 25 N. In this case, the optimization task involves determining where to place these eight boxes across the 27 candidate locations to minimize the compliance, C. To this end, an additional constraint needs to be introduced in Eq. (5.19c) so that each candidate load ($P_{j,z}$ in this case) must be equal to 25 N multiplied by an integer (from 0 to 8). This setup creates a quadratic integer programming problem that can be readily solved to achieve the global optimum. The INP optimizer, provided by MOSEK, can directly address this integer optimization (INT) problem using a branch-and-bound solver. The optimization result is displayed in red in Fig. 5.6b, c and also shown in Fig. 5.6e. Compared to the previous global optimum for the continuous variation of load magnitudes, the compliance of the new solution from the integer programming increases by only 3.61% for the more restrictive yet practical load distribution.

It is worth noting that although the finite element model for the shelf structure in Fig. 5.6a has 38,358 degrees of freedom, only a single FEA with 27 load cases needs to be performed to obtain the condensed flexibility matrix **D** and then optimize the distribution of load magnitudes. Moreover, since the number of design variables ($M = 27$) is much smaller than the degrees of freedom of the finite element model, the load distribution optimization problem can be solved very quickly using the MOC, SLSQP, and INP methods.

Next, we demonstrate how load distribution optimization can be integrated with structural topology optimization. For comparison, we consider two widely used topology optimization techniques: the BESO method (Huang and Xie 2010) and the solid isotropic material with penalization (SIMP) method (Sigmund 2001). As discussed in Sect. 4.2.3, the BESO method allows inefficient elements to be removed and efficient elements to be added simultaneously, resulting in clear 0/1 designs for elemental densities. In contrast, the SIMP method allows elemental densities to vary continuously between 0 and 1 by employing a power-law interpolation scheme to generate near 0/1 designs for elemental densities.

Figure 5.7 illustrates the computational workflow of simultaneous optimization of load condition and structural topology. In each iteration, load optimization is first performed to determine the optimized load condition for the current structure. This is immediately followed by topology optimization, which modifies the structural geometry based on the optimized load distribution. The elemental densities are iteratively updated using either BESO or SIMP based on the elemental sensitivities, which can be calculated from the global displacement vector **u** (Huang and Xie 2010; Sigmund 2001). With our single FEA approach, **u** can be obtained as

5.4 Redistributing Load Magnitudes

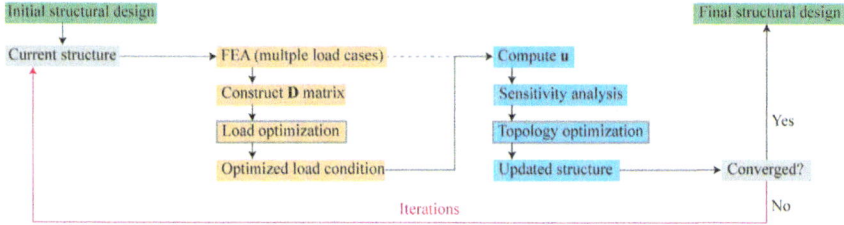

Fig. 5.7 Computational workflow of simultaneous optimization of load condition and structural topology (reprinted from Lee et al. 2025, licensed under CC-BY 4.0)

$$\mathbf{u} = \sum_{i=1}^{M} P_i \tilde{\mathbf{u}}_i \tag{5.26}$$

where $\tilde{\mathbf{u}}_i$ denotes the global displacement vector corresponding to the ith load case in the single FEA (with a unit load in each load case), and P_i is the component of the optimized load distribution. Note that only a single FEA with multiple load cases is required for each iteration shown in Fig. 5.7. This means that, even though the load conditions are different before and after the load optimization, the sensitivity information for the topology optimization in each iteration can be obtained through Eq. (5.26) without performing an additional FEA. Consequently, the total number of finite element analyses is equal to the total number of iterations.

A 2D example shown in Fig. 5.8 is used to demonstrate the simultaneous optimization of load magnitude distribution and structural topology. The structure is 210 mm wide (x direction) and 195 mm high (y direction), with a 210 mm × 15 mm rectangular non-design domain (in green). The material is assumed to be linearly elastic and isotropic, with Young's modulus of 100 GPa and Poisson's ratio of 0.30. Fixed boundary conditions are applied to both ends of the non-design domain. The entire structure is discretized using quadrilateral shell elements with a mesh size of 1.5 mm and a thickness of 1 mm. In the topology optimization process, the target volume is set to 40%. Other topology optimization parameters are $ER = 3\%$ and $r_{min} = 5$ mm.

In this example, while topology optimization aims to minimize the compliance, load optimization considers either compliance minimization ('best' load condition) or compliance maximization ('worst' load condition). These combinations result in the stiffest structures that can handle the best and worst load conditions, respectively. Specifically, the load optimization objective in this example is to distribute downward vertical loads, with a total magnitude of 500 N, across $N = 3$ candidate locations (see Fig. 5.8a). Besides, the loads are constrained between $P_{min} = -500$ N and $P_{max} = -100$ N. The MOC method is used for the load optimization.

The optimization results in Fig. 5.8b, c show four stiffest structures for the best and worst load conditions. Comparing the BESO and SIMP results, it can be observed that the two methods produce very similar outcomes in terms of compliance values

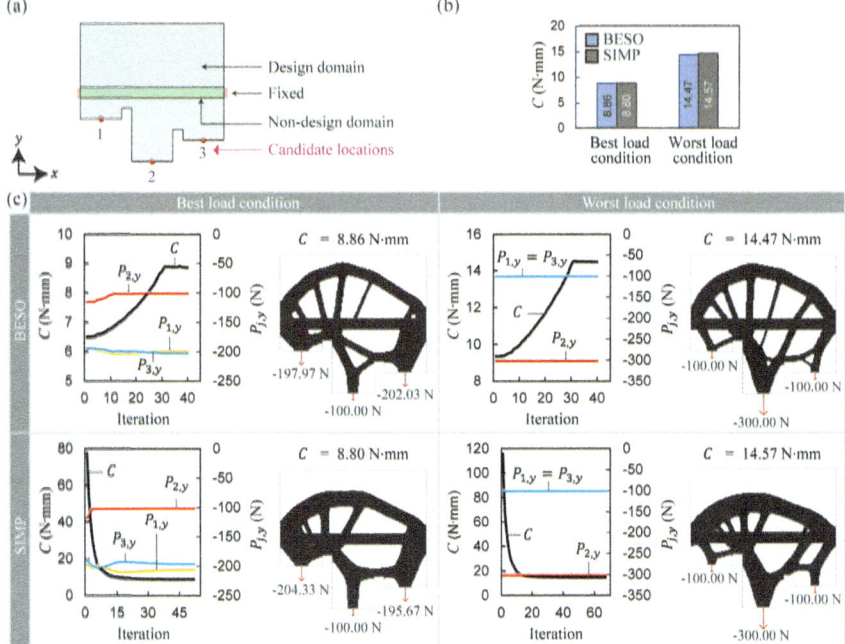

Fig. 5.8 Simultaneous optimization of load magnitude distribution and structural topology: **a** initial setup of the optimization problem; **b** comparison of compliance values of BESO and SIMP results for the best and worst load conditions; **c** detailed optimization histories and results for four cases (reprinted from Lee et al. 2025, licensed under CC-BY 4.0)

and structural topologies. For the best load condition, the optimized load distributions differ slightly between BESO and SIMP, but both methods result in location 2 reaching an optimized value of -100 N, which is P_{max}, enabling bending to be minimized. For the worst load condition, which is expected to maximize bending, it is observed that, throughout the load optimization process, the largest load (300 N) is placed at location 2 and two smallest loads (100 N each) are placed at locations 1 and 3, regardless of whether BESO or SIMP is used for the topology optimization. This example clearly demonstrates how load optimization can easily be integrated with commonly used topology optimization techniques.

5.5 Optimizing Load Directions

5.5.1 Optimization Method

Previously, Lee and Xie (2022) proposed a method for the simultaneous optimization of load direction and structural topology. One limitation of the previous approach is that all the loads had to be rotated synchronously in the same direction. Here, we present a more general method that can optimize individual directions of resultant forces at candidate locations (Lee et al. 2025). To this end, we apply a non-linear constraint on the loads, which requires that the sum of the magnitudes of individual resultant forces at candidate locations be equal to a prescribed value Q^*. Therefore, Eq. (5.17c) becomes

$$\sum_{j=1}^{N} \sqrt{(P_{j,x})^2 + (P_{j,y})^2 + (P_{j,z})^2} = Q^* \quad (5.27)$$

Notably, since the three load components $P_{j,x}$, $P_{j,y}$, and $P_{j,z}$ at every candidate location can be changed as independent design variables, Eq. (5.27) allows for the simultaneous optimization of both the direction and magnitude of the resultant force at each candidate location.

Due to the non-linear constraint in Eq. (5.27) and the non-convex nature of the optimization problem, the SLSQP method is more suitable for this case than the INP method. It should also be noted that the OC method introduced in Sect. 5.4.1 is not applicable here, as the utilized bisection algorithm requires the constraint to be monotonic for determining the Lagrange multiplier λ. Thus, the SLSQP method is adopted here and demonstrated below.

5.5.2 Numerical Examples

Figure 5.9a shows a simple 2D tree-like structure with four branches. The tree is 1,406 mm in height and 930 mm in width, with all branches being 30 mm wide. The structure is discretized using triangular shell elements with a mesh size of 3 mm and an element thickness of 1 mm. The base of the trunk is fixed, and four candidate load locations are positioned at the tips of the branches. The material is assumed to be linearly elastic and isotropic, with Young's modulus of 100 GPa and Poisson's ratio of 0.30. For this 2D structure, the eight design variables in $\mathbf{P} = [P_{1,x}, P_{1,y}, P_{2,x}, P_{2,y} P_{3,x}, P_{3,y}, P_{4,x}, P_{4,y}]^T$ can vary in both sign and magnitude, but the sum of the magnitudes of individual resultant forces at the four candidate locations must be equal to a prescribed value, Q^*, which is set to 150 N. During the optimization process, the bounds of the load components are set as $P_{\min} = -150$ N and $P_{\max} = 150$ N.

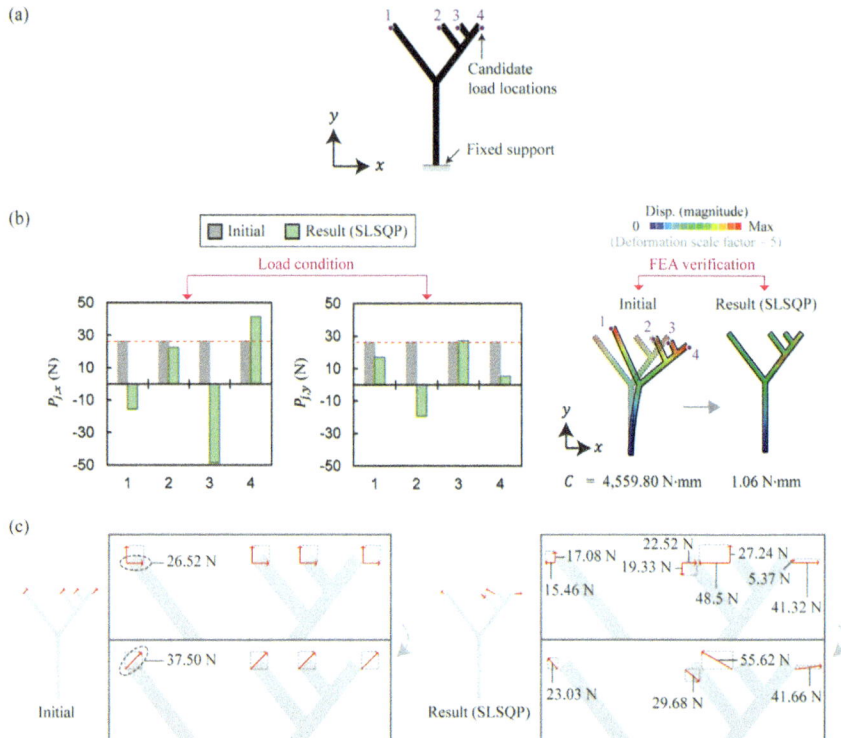

Fig. 5.9 Distributing loads to a tree-like structure, with the sum of the magnitudes of the four resultant forces meeting a prescribed value: **a** boundary condition and four candidate load locations; **b** initial and optimized load distributions (left) and FEA models verifying the results (right); **c** load components and resultant force at each candidate location for the initial (left) and optimized (right) load distributions (reprinted from Lee et al. 2025, licensed under CC-BY 4.0)

As an initial guess design, the loads are distributed equally to the four candidate locations in both x and y directions, as shown in Fig. 5.9b, c. The resultant force at each location has a magnitude of 37.5 N and forms an angle of 45° from the horizontal direction, with the total magnitude being equal to 150 N. This initial design is regarded as a feasible design as it satisfies all constraints, including Eqs. (5.27) and (5.17d). The compliance of the initial design is 4,559.80 N·mm. By applying the SLSQP method, we obtain the optimized load distribution (see Fig. 5.9b, c), which differs significantly in both direction and magnitude at each location. More importantly, the compliance corresponding to the optimized load distribution has decreased to 1.06 N·mm, representing a 98.98% reduction compared to the initial design. Such a significant improvement in the overall structural stiffness is only possible when individual load directions and magnitudes are allowed to vary so that they can reach a nearly perfect balance to minimize structural compliance, as manifested in the right-side image of Fig. 5.9b.

5.5 Optimizing Load Directions

It should be noted that since the optimization problem is non-convex here, the result from the SLSQP method cannot be guaranteed to be the global optimum. However, this should not be a major concern—as long as we can make a substantial improvement over an existing or initial design, the solution may be valuable for engineering applications. For the tree-like structure shown in Fig. 5.9a, we have conducted extensive numerical tests by applying the SLSQP method to 400 randomly generated feasible initial designs of load distributions, examining the compliance values before and after optimization (Lee et al. 2025). We find that all results show a substantial reduction in the compliance, with an average decrease of 99.61%. Indeed, many of these 'nearly-optimal' solutions could be practically useful as diverse and competitive designs.

In the next example, we demonstrate a potential engineering application of compliance maximization through load direction optimization. Specifically, we aim to generate the maximum amount of elastic energy in a curved-folded bow shown in Fig. 5.10a. To achieve this goal, we should maximize the compliance rather than designing for the stiffest structure. Due to symmetry, only half of the bow needs to be modelled, reducing the candidate load locations to just one—at the intersection of the bow and the string.

The bounding dimensions of the half bow are approximately 328 mm × 114 mm × 40 mm in the x, y, and z directions, respectively. The half bow model is discretized using triangular shell elements with a mesh size of 5 mm and an element thickness

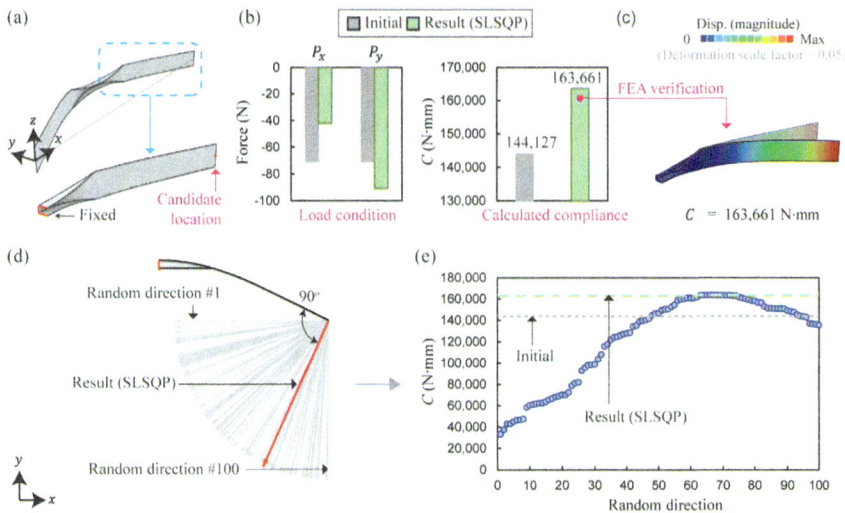

Fig. 5.10 Generating the maximum amount of elastic energy in a curved-folded bow: **a** initial setup of the optimization problem; **b** load conditions and compliance values before and after optimization; **c** performing an additional FEA to verify the optimization result; **d** optimized load direction (in red) and 100 randomly chosen feasible load directions; **e** comparison of compliance values corresponding to the optimized and random load directions (reprinted from Lee et al. 2025, licensed under CC-BY 4.0)

of 1 mm. The material properties are set the same as in the previous example. The load is assumed to act only in the xy plane only. Therefore, the optimization involves two load components, P_x and P_y. The target magnitude of the resultant force, Q^* in Eq. (5.27), is set to 100 N. Besides, the load components are constrained between $P_{\min} = -100$ N and $P_{\max} = 0$ N.

As a feasible initial design, we select $P_x = P_y = -70.71$ N, which generates a compliance of 144,127 N·mm. The optimization results from the SLSQP method are given in Fig. 5.10b, c, showing that the maximized compliance reaches 163,661 N·mm—representing a 13.55% increase compared to the initial design. Interestingly, as shown in Fig. 5.10d, the optimized resultant load is oriented perpendicular to the bow, which is reasonable as it maximizes bending.

To further verify the optimization result, we apply the SLSQP method starting from 100 feasible initial designs generated from randomly chosen load directions (see Fig. 5.10d). Figure 5.10e reveals that the optimized load direction achieves the highest compliance among all load directions. This example further demonstrates the effectiveness of our method for optimizing load directions. Moreover, this example suggests an alternative application of compliance maximization: rather than preventing structural failure due to excessive deformation, it can be used to generate the largest possible elastic energy in a structure, which could then be released to achieve a desired dynamic response.

5.6 Conclusion

In this chapter, we have demonstrated that structural performance can be significantly enhanced by optimizing load distributions, including the locations, magnitudes, and directions of the loads. We have developed several methods for solving load distribution optimization problems. Specifically, the optimality criteria (OC) method and the modified optimality criteria (MOC) method can be used to optimize load locations and load magnitudes, respectively; the interior-point (INP) method is capable of finding the globally optimal distribution of load magnitudes for compliance minimization problems; and the sequential least squares quadratic programming (SLSQP) method can be employed to optimize the load magnitudes and directions for both compliance minimization and compliance maximization problems. Additionally, we have presented an efficient method for finding the globally optimal load locations by performing a single FEA with multiple load cases. Further, we have demonstrated that these load distribution optimization methods can be integrated with commonly used topology optimization techniques, such as BESO and SIMP, to create highly innovative and efficient structural designs that are unattainable through conventional topology optimization with predetermined load conditions.

References

Cross, R. (1998) The sweet spot of a baseball bat. *Am. J. Phys.* **66**, 772–779.

Goulart, P. J. and Chen, Y. (2024) Clarabel: An interior-point solver for conic programs with quadratic objectives. arXiv:2405.12762.

Huang, X. and Xie, Y. M. (2010) *Evolutionary Topology Optimization of Continuum Structures: Methods and Applications.* Chichester: John Wiley & Sons.

Jeong, S., Seong, H. K., Kim, C. W. and Yoo, J. (2019) Structural design considering the uncertainty of load positions using the phase field design method. *Finite Elem. Anal. Des.* **161**, 1–15.

Kraft D. (1988) A software package for sequential quadratic programming. Tech. Rep. DFVLR-FB 88–28, German Aerospace Centre.

Lee, T.-U., Lu, H. and Xie, Y. M. (2025) Optimizing load distributions in structural design. *Eng. Struct.* (to appear).

Lee, T.-U. and Xie, Y. M. (2022) Optimizing load locations and directions in structural design. *Finite Elem. Anal. Des.* **209**, 103811.

Lee, T.-U. and Xie, Y. M. (2023) Finding globally optimal arrangements of multiple point loads in structural design using a single FEA. *Struct. Multidisc. Optim.* **66**, 90.

MOSEK (2019) MOSEK optimization toolbox for MATLAB 9.0.105. http://docs.mosek.com/9.0/toolbox/index.html. Accessed 8 December 2024.

Nesterov Y. and Nemirovskii A. (1994) *Interior-Point Polynomial Algorithms in Convex Programming.* Philadelphia: Society for Industrial and Applied Mathematics.

SciPy (2024) scipy.optimize.minimize. https://docs.scipy.org/doc/scipy/reference/generated/scipy.optimize.minimize.html. Accessed 8 December 2024.

Sigmund, O. (2001) A 99 line topology optimization code written in Matlab. *Struct. Multidisc. Optim.* **21**, 120–127.

Sorkine, O., Cohen-Or, D., Lipman, Y., Alexa, M., Rössl, C. and Seidel, H.-P. (2004) Laplacian surface editing. *Proc. 2004 Eurographics/ACM SIGGRAPH Symp. Geometry Process.*, Nice, 8–10 July 2004, 175–184.

Takezawa, A., Nii, S., Kitamura, M. and Kogiso, N. (2011) Topology optimization for worst load conditions based on the eigenvalue analysis of an aggregated linear system. *Comput. Methods Appl. Mech. Eng.* **200**, 2268–2281.

Xie, Y. M. and Steven, G. P. (1997) *Evolutionary Structural Optimization.* London: Springer.

Open Access This chapter is licensed under the terms of the Creative Commons Attribution 4.0 International License (http://creativecommons.org/licenses/by/4.0/), which permits use, sharing, adaptation, distribution and reproduction in any medium or format, as long as you give appropriate credit to the original author(s) and the source, provide a link to the Creative Commons license and indicate if changes were made.

The images or other third party material in this chapter are included in the chapter's Creative Commons license, unless indicated otherwise in a credit line to the material. If material is not included in the chapter's Creative Commons license and your intended use is not permitted by statutory regulation or exceeds the permitted use, you will need to obtain permission directly from the copyright holder.

Chapter 6
Human–Computer Interaction

This chapter presents an interactive topology optimization method that incorporates the designer's subjective preferences. We have developed two techniques: one based on drawing an initial pattern and the other on assigning subjective scores to intermediate designs. These techniques use the designer's initial pattern or subjective scores to modify the sensitivity values of elements, thereby altering their relative ranking. This leads to innovative and efficient structural designs that account for the designer's subjective preferences. Further, this chapter introduces our recent research on integrating virtual reality (VR) technology with topology optimization. We demonstrate that VR sculpting offers the designer an interactive, intuitive, and immersive platform for visualizing and editing 3D geometries. The sculpted 3D models can effectively incorporate the designer's subjective preferences and influence material redistribution in the generalized topology optimization process through human–computer interaction.

6.1 Introduction

Topology optimization has undoubtedly become a powerful tool for finding high-performance structural designs. However, in real-world projects, structural performance is only one of many factors that must be considered. For example, when applying topology optimization in architectural design projects, the designer often needs to balance structural efficiency with aesthetic preferences and functional requirements. A major challenge is that these preferences are usually subjective, undetermined, and evolving. To address this challenge, we have developed a human–computer interaction platform that allows the designer to express their preferences *at the outset* or *during* the design exploration process. Rather than treating the topology optimization tool as a 'black box' and passively waiting for it to produce a single solution that is often unsatisfactory or disappointing, the new platform empowers the designer to actively control or influence the optimization process by providing an initial pattern or assigning subjective scores to intermediate designs. Further, we have integrated VR technology into the interactive platform to enable immersive visualization and intuitive design exploration of 3D structures. Much of the material presented in this chapter is based on the work of Li et al. (2023a, 2024a, 2025).

It is worth noting that while the bi-directional evolutionary structural optimization (BESO) method is adopted in this chapter, the human–computer interaction techniques developed here can also be applied to other topology optimization methods.

6.2 Indicating Preferences by Drawing a Pattern

The first technique we have developed to incorporate subjective preferences in topology optimization allows the designer to draw a pattern, as shown in Fig. 6.1a. This enables the designer to indicate their creative ideas based on artistic intuition. However, a design produced directly from such an intuitive pattern may be highly inefficient in term of structural performance. Therefore, a balance should be found between the preferred pattern and structural efficiency. This can be achieved by using the drawn pattern to alter the sensitivity numbers of individual elements, thereby guiding the topology optimization process to produce designs that account for both structural performance and subjective preference. Unlike the non-design domain introduced in Sect. 3.3, the drawn pattern here can be altered during the topology optimization process.

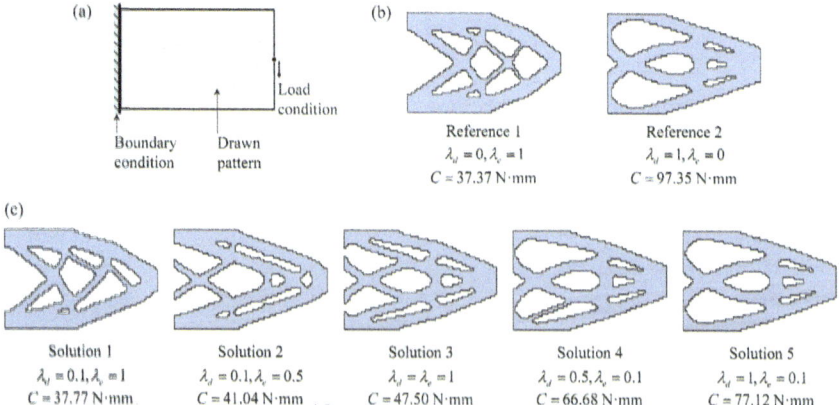

Fig. 6.1 Using a drawn pattern to influence the design outcome of a short cantilever: **a** the drawn pattern, and boundary and load conditions; **b** extreme designs from considering structural performance only (Reference 1) and subjective preference only (Reference 2); **c** five design solutions obtained from using different combinations of λ_d and λ_e (reprinted from Li et al. 2023a, with permission from Elsevier)

6.2 Indicating Preferences by Drawing a Pattern

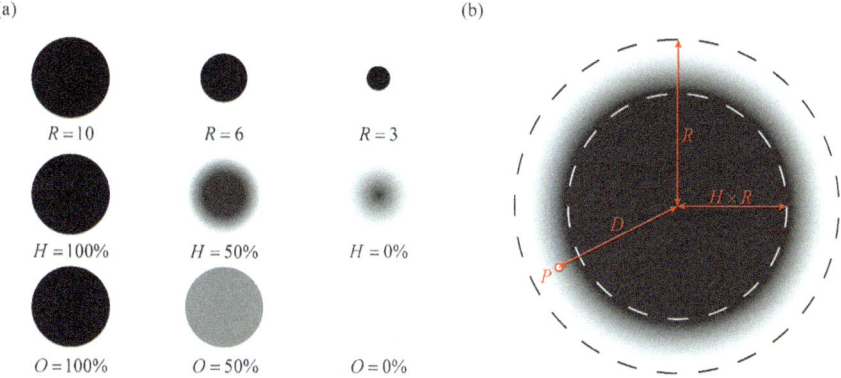

Fig. 6.2 Creating a brush stroke: **a** effects of varying brush parameters; **b** diagram explaining parameters of a stroke (reprinted from Li et al. 2023a, with permission from Elsevier)

A brush tool is developed to enable the designer to draw a pattern using a computer mouse by clicking and swiping. With each click, a circular stroke is created at the position of the cursor. Inspired by the popular image processing tool, Adobe Photoshop, three parameters are defined to produce rich drawing effects of the stroke: radius (R), hardness (H), and opacity(O). H and O range from 0 to 100%, with H controlling the sharpness of the stroke's edge and O adjusting the shade of grey of the stroke (see Fig. 6.2a).

Figure 6.2b shows a point P in a stroke. Its shade of grey is calculated as

$$s_P = \begin{cases} O, & \text{if } D \leq RH \\ O\frac{R-D}{R-RH}, & \text{if } RH < D < R \\ 0, & \text{if } D > R \end{cases} \quad (6.1)$$

where D is the distance between point P and the centre of the stroke, and s_P ranges from 0 (white) to 1 (black). To simulate the effect of manual drawing, a series of closely spaced strokes is created by swiping the cursor, with new strokes overlapping the existing ones. For any overlapping point, the highest s_P value obtained from Eq. (6.1) is taken as the final shade of grey for point P. As a result, the drawn pattern is converted into a greyscale raster texture composed of pixels of varying shades of grey. The drawing weight of the ith element, ω_i^d, is calculated as

$$\omega_i^d = \frac{\sum_{j=1}^{n} s_j}{n}, \ s_j \in [0, 1] \quad (6.2)$$

where n is the number of nodes in the ith element and s_j is the shade of grey of the jth node, calculated from Eq. (6.1).

As discussed in Sect. 4.2.3, the contribution of an element to the structural stiffness can be estimated by its sensitivity number $\tilde{\alpha}_i$, calculated from Eq. (4.14). Here, the normalized elemental sensitivity number, ω_i^e, is used:

$$\omega_i^e = \frac{\tilde{\alpha}_i - \alpha_{\min}}{\alpha_{\max} - \alpha_{\min}} \tag{6.3}$$

where $\alpha_{\min}$ and $\alpha_{\max}$ are the minimum and maximum values of $\tilde{\alpha}_i$, respectively.

To achieve the desired effect of the drawn pattern, a modified sensitivity number, $\overline{\alpha}_i^d$, is employed in the topology optimization process

$$\overline{\alpha}_i^d = \lambda_d \omega_i^d + \lambda_e \omega_i^e, \ \lambda_d, \lambda_e \in [0, 1] \tag{6.4}$$

where λ_d and λ_e are parameters ranging from 0 to 1, which indicate the relative importance of the drawn pattern (represented by ω_i^d) and the structural performance (represented by ω_i^e), respectively. For a performance-driven design (e.g., a bridge), λ_d should be small (close to 0) and λ_e large (close to 1). Conversely, for a preference-driven design (e.g., a chair), λ_d should be large (close to 1) and λ_e small (close to 0).

Figure 6.1 illustrates the effect of using different combinations of λ_d and λ_e on the design outcome of a short cantilever, starting from a preferred pattern drawn with the brush tool (see Fig. 6.1a). This pattern may reflect the designer's aesthetic preferences or be based on an existing design. Figure 6.1b shows two extreme designs: one considering only structural performance (Reference 1, $\lambda_d = 0$ and $\lambda_e = 1$) and the other considering only subjective preference (Reference 2, $\lambda_d = 1$ and $\lambda_e = 0$). As shown in Fig. 6.1c and Table 6.1, when λ_d gradually increases and λ_e decreases, we obtain five different designs, with diminishing structural efficiency but growing resemblance to the drawn pattern.

The similarity between two geometries can be measured in many different ways. A simple measure is the overlapping ratio Q (Wang et al. 2018), defined as

$$Q = \frac{V_Q}{V^*} \tag{6.5}$$

where V_Q represents the overlapping regions between two designs with the same total volume, V^*. For example, there is an overlapping ratio of 89% between Solutions A and B in Fig. 6.3. However, relying solely on the overlapping ratio may not always capture the topological differences between two solutions. A simple yet informative metric for assessing topological characteristics is the number of internal holes, g, also known as the genus. For example, Solution A has a genus of 3, while Solution B exhibits a genus of 5.

In Table 6.1, we provide a detailed comparison of both structural performance and geometric characteristics of all the designs for the short cantilever shown in Fig. 6.1. The structural performance is measured by C/C_1, which is the relative compliance

6.2 Indicating Preferences by Drawing a Pattern

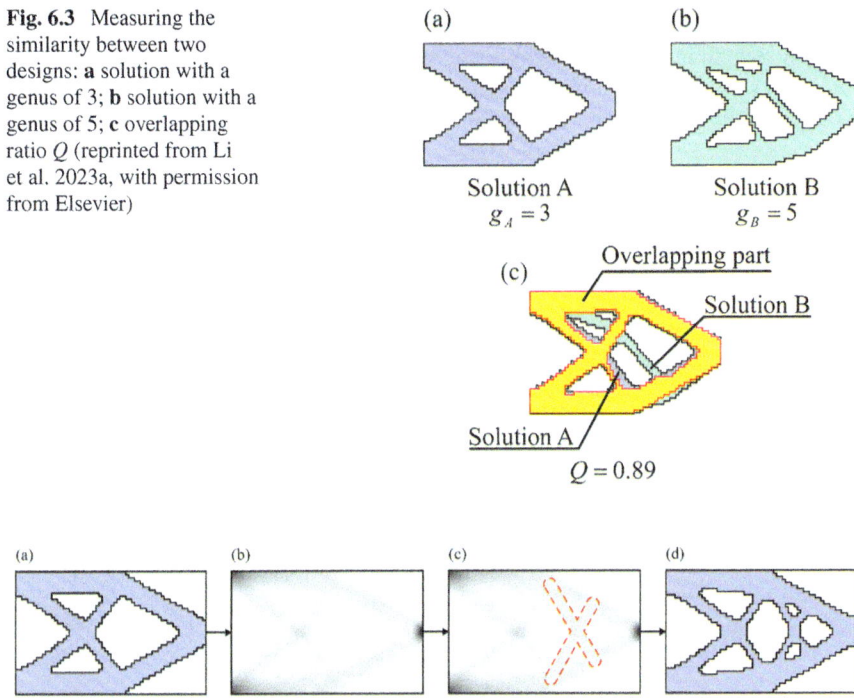

Fig. 6.3 Measuring the similarity between two designs: **a** solution with a genus of 3; **b** solution with a genus of 5; **c** overlapping ratio Q (reprinted from Li et al. 2023a, with permission from Elsevier)

Fig. 6.4 Creating a new pattern based on a previous solution: **a** parent design; **b** nodal sensitivity field; **c** drawing additional features; **d** child design (reprinted from Li et al. 2023a, with permission from Elsevier)

with respect to Reference 1. The geometrical characteristics are measured by the genus, g, and Q_2 (the overlapping ratio between each solution and Reference 2).

Note that in the example shown in Fig. 6.1, the drawn pattern is provided at the outset of the topology optimization process. To enhance interactions between human and computer, we also allow the designer to modify a pattern based on a previous topology optimization solution, which is referred to as a 'parent design' (see Fig. 6.4a). To use the parent design in the subsequent topology optimization, its nodal sensitivity field is first converted into a texture (see Fig. 6.4b) using a bi-linear interpolation method (Kirkland, 2010). The designer can then add new features to,

Table 6.1 Comparison of structural performance and geometrical characteristics of different designs for the short cantilever shown in Fig. 6.1

	Reference 1	Reference 2	Solution 1	Solution 2	Solution 3	Solution 4	Solution 5
C/C_1	1	2.61	1.01	1.10	1.27	1.78	2.06
g	8	6	7	7	6	7	6
Q_2	0.59	1	0.65	0.77	0.84	0.94	0.97

or delete existing parts from, the parent design. Starting from the revised pattern in Fig. 6.4c, which integrates the additional drawing and the parent design, a new cycle of topology optimization is performed to generate the 'child design' (see Fig. 6.4d). After this, if the designer is still not fully satisfied with the outcome, they can repeat the process for several more generations, using the current child design as the new parent design. This iterative and interactive process enables the designer to adjust their drawn pattern adaptively so that a desired balance between subjective preferences and structural performance can be achieved eventually.

6.3 Assigning Scores to Intermediate Designs

The second technique we have developed for incorporating subjective preferences in topology optimization is to assign scores to intermediate designs—the parent designs for the next generation of child designs. The designer can give each parent design a score, S, based on their artistic preferences or other considerations. These subjective scores are then used to modify the elemental sensitivity numbers and influence the outcome of the next round of topology optimization. Specifically, the raw sensitivity number $\tilde{\alpha}_i$ from Eq. (4.14) is changed to

$$\overline{\alpha}_i^d = \omega_i^s \tilde{\alpha}_i \tag{6.6}$$

where $\overline{\alpha}_i^d$ is the modified sensitivity number of the ith element and ω_i^s is the scoring weight defined as

$$\omega_i^s = 1 + \lambda_s \rho_i \frac{S}{10}, \quad \lambda_s \in [0, 1], \quad S \in [-5, 5], \quad \rho_i = 1 \text{ or } \rho_{\min} \tag{6.7}$$

in which λ_s is a parameter ranging from 0 to 1, which allows the designer to control the level of influence of the score on topology optimization, and ρ_i is the density of the ith element ($\rho_i = 1$ for solid element, and $\rho_i = \rho_{\min}$ for void element). The subjective score S can be any value from -5 to 5. Together, the scoring weight ω_i^s varies within the range of [0.5, 1.5].

We illustrate the effect of applying subjective scores using a simple example in Fig. 6.5, assuming $\lambda_s = 1$. It can be observed that applying a negative score ($S = -2$) and a positive score ($S = 2$) to the same parent design results in significantly different child designs. When $S = -2$, the parent design is regarded as an undesirable solution. As a result, the subsequent topology optimization tries to avoid producing a solution similar to the parent design, leading to a distinctly different child design for the designer to consider. In contrast, when $S = 2$, the parent design is regarded as a favourable solution, and the child design ends up only slightly different from the parent design. Note that setting S to 0 will result in no change to the parent design in the subsequent topology optimization.

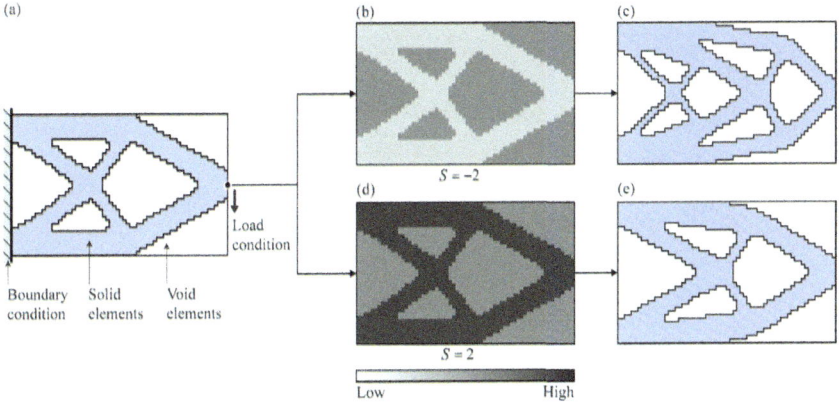

Fig. 6.5 Applying subjective scores in topology optimization: **a** parent design; **b** when $S = -2$, the scoring weight ω_i^s ranging from 0.8 to 1; **c** child design obtained by assigning $S = -2$; **d** when $S = 2$, the scoring weight ω_i^s ranging from 1 to 1.2; **e** child design obtained by assigning $S = 2$ (reprinted from Li et al. 2023a, with permission from Elsevier)

6.4 Combining Drawing and Scoring in a Multi-solution System

To provide more options to the designer and accelerate the design exploration process, we have incorporated a multi-solution technique in the human–computer interaction platform. The multi-solution technique is based on the random perturbation method introduced in Sect. 2.6.2. By integrating the multi-solution technique with the subjective drawing and scoring algorithms, we obtain the following combined sensitivity number from Eqs. (2.1), (6.4), and (6.6):

$$\overline{\alpha}_i = \beta_i \omega_i^s \left(\lambda_d \omega_i^d + \lambda_e \omega_i^e \right), \; \lambda_d, \lambda_e \in [0, 1] \tag{6.8}$$

where β_i is a penalty coefficient randomly selected from a certain range, e.g., [0.85, 1.15]. Note that the random coefficients of all elements are generated prior to the optimization process, and the generation of these random coefficients is driven by a random seed integer, γ. Different values of γ produce different sets of random coefficients. When $\gamma = 0$, β_i is equal to 1 and the optimization solver is not influenced by the penalty coefficient.

Based on Eq. (6.8), we have developed a modified BESO method that incorporates subjective preference (SP). We call it the SP-BESO method. The computational workflow of the SP-BESO method is summarized in Fig. 6.6. This method is adapted from the soft-kill BESO method presented in Sect. 4.2.3. Note that the SP-BESO method does not increase the algorithmic complexity of the underlying BESO method, as it simply uses a modified sensitivity number for each element based on Eq. (6.8).

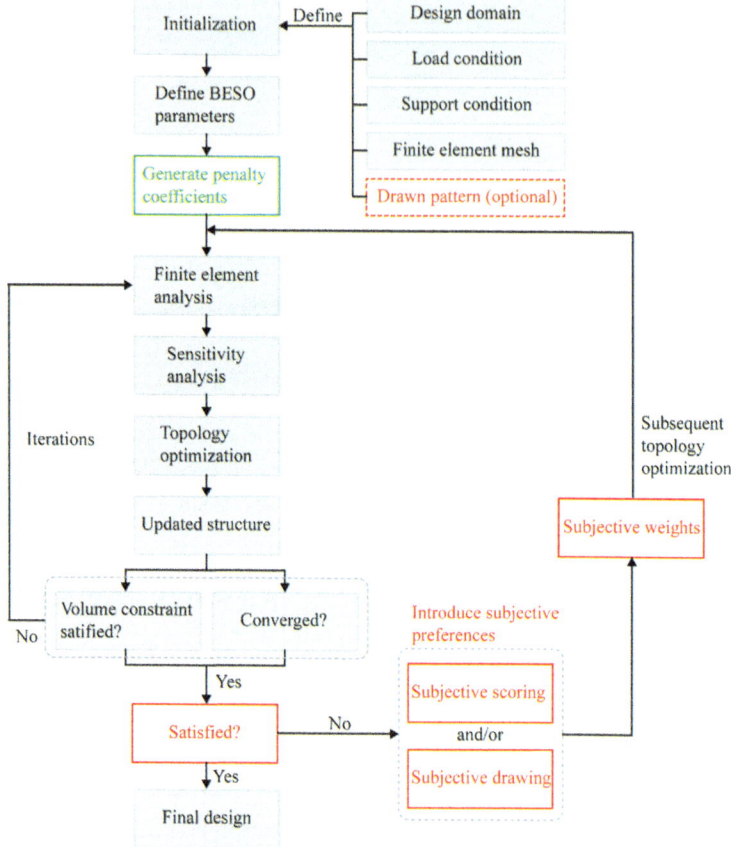

Fig. 6.6 Computational workflow of the SP-BESO method. Note that in the multi-solution system, topology optimization is performed on multiple parent designs simultaneously (in parallel) (reprinted from Li et al. 2023a, with permission from Elsevier)

We have implemented the combined system in an interactive digital design tool called iBESO. This software, along with a user guide, has been made publicly available (Li et al. 2023b). All examples presented in Sects. 6.2–6.4 are produced using the iBESO software.

As shown in Fig. 6.7, iBESO can simultaneously run up to four 2D topology optimization solvers, each with a different set of penalty coefficients β_i. This allows the designer to evaluate and modify four different parent designs in parallel on the same screen through drawing and scoring, which leads to the next generation of four child designs in the subsequent topology optimization. The cycle of evaluation, modification, and optimization can be repeated multiple times until a satisfactory design is obtained.

6.4 Combining Drawing and Scoring in a Multi-solution System

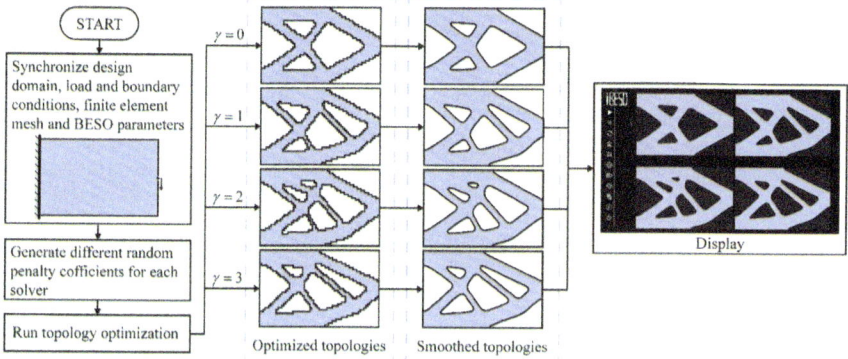

Fig. 6.7 Computational workflow of the multi-solution system (reprinted from Li et al. 2023a, with permission from Elsevier)

To enhance the aesthetic quality of the optimization results, the zigzagged boundaries in the finite element models are usually smoothed (Li et al. 2022) before being displayed on the screen for the designer to evaluate and modify (see Fig. 6.7).

We demonstrate the capability of the SP-BESO method through a design exploration of a chair. Unlike engineering structures such as a bridge, the load-carrying capacity of furniture such as a chair plays a secondary role in its design. Nevertheless, there is growing interest in using topology optimization to generate innovative designs and organic forms for furniture (Ma et al. 2021). The SP-BESO method enables the designer to interact with the computer by controlling or influencing the topology optimization process to achieve a desired outcome.

Figure 6.8a illustrates a 2D design space of 800 mm × 800 mm for a chair. The seat and backrest areas are set as a continuous non-design domain, subjected to a uniformly distributed load of 1 N/mm². Two points at the bottom of the design space are fixed. A row of passive void elements is specified above the seat and along the left-hand side of the backrest to prevent material from being added in these areas. The BESO parameters used in this example are $V^* = 20\%$, $r_{min} = 30$ mm, and $ER = 3\%$. Here, the goal is to create a series of preference-driven designs that exhibit reasonable structural performance.

First, three reference designs are obtained, as shown in Fig. 6.8a. Reference 1 is generated using the conventional BESO method, while References 2 and 3 are derived from using two different drawn patterns in SP-BESO with $\lambda_e = 0$, $\lambda_s = 0$, and $\lambda_d = 1$. The drawn pattern of Reference 2 is inspired by the Wiggle Side Chair originally designed by the renowned architect Frank Gehry (Martin 2018). Reference 3 is generated from a drawn pattern that provides no direct support to the seating area. This is intentionally chosen to represent a highly inefficient structural design, possibly created by an artist who is unfamiliar with engineering principles. Obviously, Reference 1 is a performance-driven solution, whereas References 2 and 3 are preference-driven designs. The compliance values of References 2 and 3 are 16.6 times and 27.6 times higher than that of Reference 1, respectively.

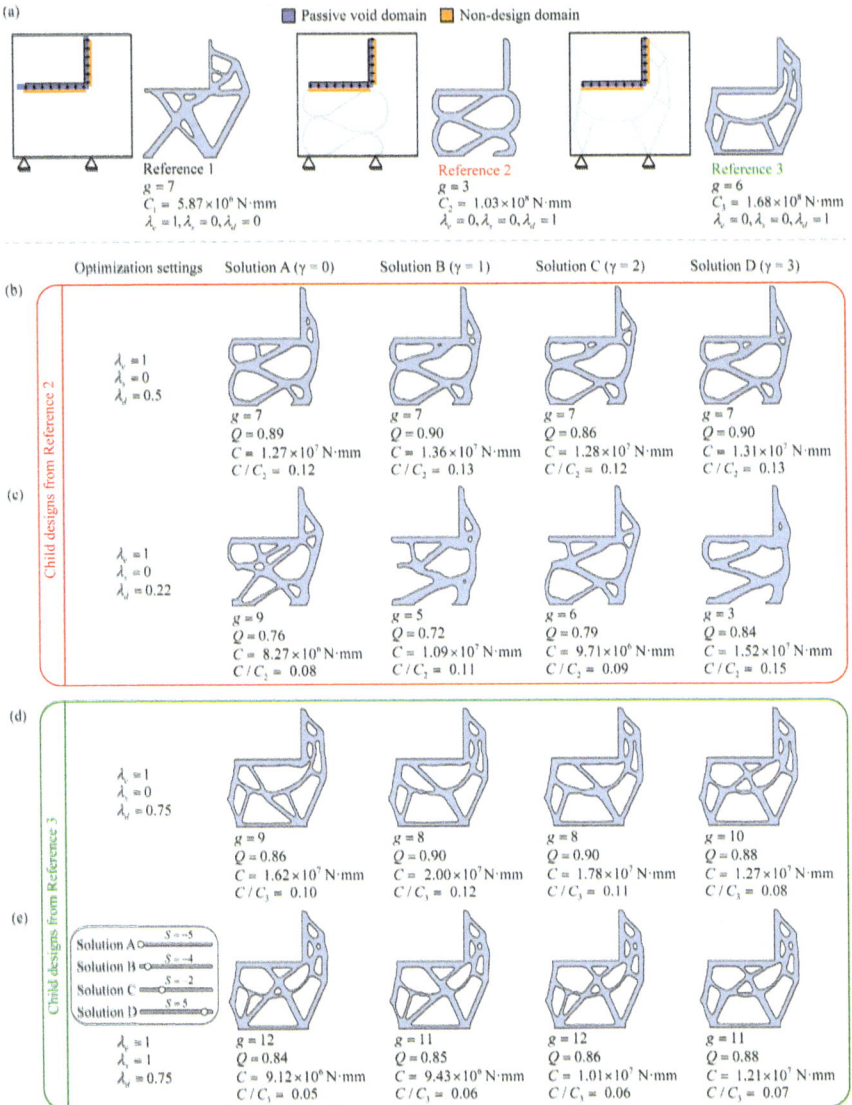

Fig. 6.8 Design exploration of a chair: **a** Reference 1 from conventional BESO, and References 2 and 3 from using different drawn patterns in SP-BESO; **b** child designs from Reference 2 when $\lambda_d = 0.5$; **c** child designs from Reference 2 when $\lambda_d = 0.22$; **d** child designs from Reference 3 when $\lambda_d = 0.75$; **e** a new generation of child designs after assigning subjective scores to the previous generation (reprinted from Li et al. 2023a, with permission from Elsevier)

6.4 Combining Drawing and Scoring in a Multi-solution System

In this example, we employ two different strategies to generate child designs: (1) only adjusting the weighting parameter for drawing, λ_d; and (2) combining the initial drawing and subsequent scoring over two generations. First, we apply strategy (1) to Reference 2, using four sets of random penalty coefficients β_i, generated by setting the random seed integer, γ, to 0, 1, 2, and 3, respectively. The first four child designs (see Fig. 6.8b) are obtained using $\lambda_e = 1$, $\lambda_s = 0$, and $\lambda_d = 0.5$, aiming to achieve improved structural performance with small distortions to the drawn pattern. Interestingly, it can be clearly observed that the four child designs have successfully preserved the main features of the drawn pattern and have added new members, especially behind the backrest, to enhance the structural performance. Although the genus number, g, has increased from 3 to 7 due to the addition of four small holes, the overlapping ratio, Q, remains high, ranging from 86 to 90%. Most significantly, the compliance, C, has been reduced by 88%, 87%, 88%, and 87% in Solutions A–D, respectively.

If we allow larger distortions to the drawn pattern by selecting a smaller λ_d of 0.22, we obtain a new set of four child designs, of which three exhibit even better structural performance (see Fig. 6.8c). The best child design, Solution A, has its compliance reduced by 92% compared to Reference 2. Remarkably, the compliance of this child design is only 41% higher than that of Reference 1, even though the compliance of the parent design (Reference 2) is 16.6 times greater than that of Reference 1. It is also noted that with this improved structural performance, the overlapping ratio with respect to Reference 2 has decreased to 76%, 72%, 79%, and 84% in Solutions A–D, respectively. Additionally, the genus number varies significantly, ranging from 9 in Solution A to 3 in Solution D.

We apply strategy (2) to Reference 3. The first set of child designs are generated using $\lambda_e = 1$, $\lambda_s = 0$, and $\lambda_d = 0.75$ (see Fig. 6.8d). It is observed that the four child designs have preserved key geometric features of the drawn pattern and added new members to support the seat and backrest, resulting in significantly improved structural performance. The compliance is reduced by 90%, 88%, 89%, and 92% in Solutions A–D, respectively. The overlapping ratio is high, ranging from 86 to 90%. The genus number varies between 8 and 10, compared to 6 in Reference 3.

After evaluating the geometric features and structural performance of the four child designs in Fig. 6.8d, the designer may assign each design a score, S, based on their subjective preferences. For example, in this case, we assign scores of $S = -5$, -4, -2, and 5 to Solutions A–D, respectively, indicating that Solution D is currently the most favourable design. We then perform SP-BESO again by setting $\lambda_e = 1$, $\lambda_s = 1$, and $\lambda_d = 0.75$, which leads to the second generation of child designs in Fig. 6.8e. Interestingly, all four new child designs have inherited most of the geometric features of the drawn pattern in their 'grandparent', Reference 3. Furthermore, each new child design has achieved better structural performance than its own parent.

Note that in our current SP-BESO algorithm, each child design is generated from a single parent. However, it is well known that in both natural and artificial systems, crossover (recombination) between two or more parents can be beneficial for producing stronger offspring (Holland, 1975). Further research along this line would be interesting and valuable, particularly by building on the techniques for creating diverse and competitive designs discussed in Chaps. 2–5.

6.5 Performing Interactive Topology Optimization in Virtual Reality

So far, our discussion on establishing an interactive topology optimization platform has been limited to 2D problems, using a computer mouse. Its extension to 3D problems might seem straightforward but actually faces major challenges. A significant barrier is the inherent difficulty in visualizing and modifying 3D models. A designer who lacks strong spatial awareness often struggles to interpret and interact with 3D models on a 2D computer screen (Wolfartsberger 2019). In addition, results from 3D topology optimization typically include internal geometric details that are not visible or editable from the outside, which further complicates the process (Li et al. 2023c).

To address these challenges, we have developed a new design exploration platform that integrates VR technology and the SP-BESO method (Li et al. 2024a). This integration provides a virtual design environment where the designer can intuitively interact with and edit complex 3D geometries. In this environment, the designer can first express their subjective preferences by sculpting (or importing) a preferred model or pattern. This pattern is then translated into weights to guide the material redistribution in a modified topology optimization method for creating a 3D structure, incorporating the designer's preferences expressed through the sculpted pattern. The designer can further refine their subjective preferences by manually adding or removing parts using VR sculpting tools. This enables the designer to efficiently explore multiple design options through iteratively updating subjective preferences and executing optimization.

We use a head-mounted display and two handheld controllers of the Meta Quest 3 VR device to visualize and edit 3D geometries (Meta 2024), as shown in Fig. 6.9a. Such a VR device enables the user to perform operations based on body motion, offering an intuitive way to interact with the virtual environment. The immersive nature of this environment is particularly beneficial for inexperienced designers, as it helps them better understand complex 3D details and conveniently refine their subjective preferences. The VR-based BESO technique has been implemented in an interactive digital design tool called VR-BESO. This software and its user guide have been made publicly available (Li et al. 2024b). The VR-BESO software has been used to produce all examples presented in Sect. 6.5.

There are four main steps involved in a single design exploration cycle of the interactive topology optimization workflow (see Figs. 6.9 and 6.10). In Step 1 (see Fig. 6.9a), the designer uses VR sculpting to draw a 3D model in the virtual environment by waving the two handheld controllers. This process provides a direct and intuitive way for the designer to express their subjective preferences, making the process easily accessible even to those without 3D modelling expertise. VR sculpting

6.5 Performing Interactive Topology Optimization in Virtual Reality   119

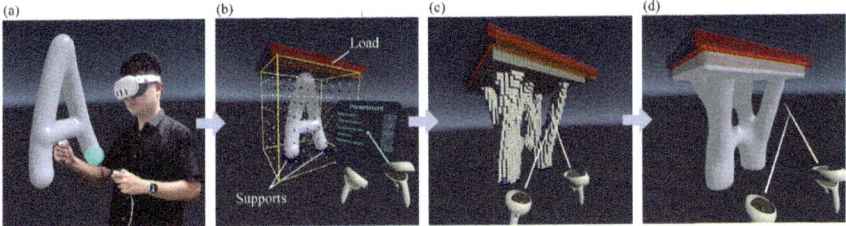

Fig. 6.9 Main steps in one design exploration cycle: **a** Step 1: VR sculpting; **b** Step 2: initialization; **c** Step 3: topology optimization; **d** Step 4: smoothing (reprinted from Li et al. 2024a, with permission from authors)

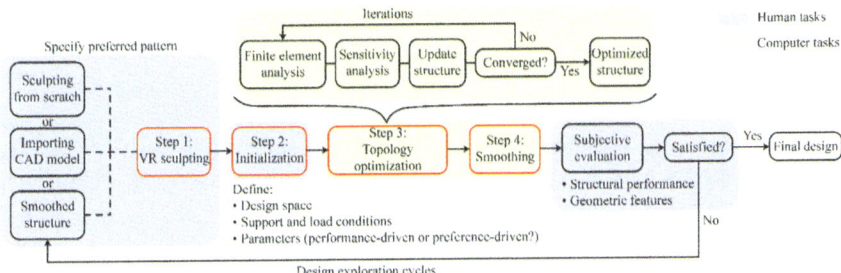

Fig. 6.10 Computational workflow of the interactive design exploration platform using VR (reprinted from Li et al. 2024a, with permission from authors)

differs significantly from the traditional 2D-screen user interface, offering vast design freedom and simulating real-life sculpting experience (Chen et al. 2021). The VR technology typically includes various brush tools for drawing, adding, and removing material.

Step 1 also allows the designer to import a preexisting model, a smoothed solution from a previous design exploration cycle, or a 3D geometry created using other computer-aided design (CAD) software, e.g., Rhino (Robert McNeel & Associates 2024). The imported design is first converted into an editable model using a surface reconstruction technique (Kazhdan et al. 2006). The designer can then use VR sculpting tools to add new features or remove unwanted parts, guided by their artistic intuition or preferences. This capability may save significant time that would otherwise be spent sculpting a model from scratch, enabling the designer to focus on refining their subjective preferences.

In Step 2 (see Fig. 6.9b), the designer begins by specifying the dimensions of the design space using the handheld controllers to adjust the cuboidal bounding box. Next, the support and load conditions can be defined. For instance, a uniformly distributed load, represented by a red rectangular box in Fig. 6.9b, is applied to the top surface of the non-design domain located in the upper region of the design space, while fixed supports are applied at the base. Finally, topology optimization parameters, such as the filter radius and the volume fraction, are specified through a virtual number pad.

In Step 3 (see Fig. 6.9c), a modified topology optimization process based on the BESO method is performed. First, the sculpted pattern (or imported model), Ω, is converted into subjective weights by computing a distance field

$$d_i = \begin{cases} d(c_i, \partial\Omega), & \text{if } c_i \in \Omega \\ 0, & \text{if } c_i \notin \Omega \end{cases} \quad (6.9)$$

where c_i is the centroid of the ith element, $\partial\Omega$ represents the boundary of Ω, and $d(c_i, \partial\Omega)$ is the shortest distance between c_i and $\partial\Omega$. As a relative measure, the normalized distance, d_i^n, is used

$$d_i^n = \frac{d_i - d_{\min}}{d_{\max} - d_{\min}} \quad (6.10)$$

where $d_{\min}$ and $d_{\max}$ are the minimum and maximum values of d_i, respectively. The revised sensitivity number, $\bar{\alpha}_i$, that is used in the modified topology optimization method is defined as

$$\bar{\alpha}_i = \lambda d_i^n + (1 - \lambda)\omega_i^e, \ \lambda \in [0, 1] \quad (6.11)$$

where ω_i^e is the normalized elemental sensitivity number given in Eq. (6.3) and λ is a parameter ranging from 0 to 1 that controls the level of influence of the sculpted pattern. When $\lambda = 0$, the sculpted pattern has no effect on the topology optimization process; when $\lambda = 1$, the sculpted pattern dictates the final shape. To achieve a balance between structural performance and the subjective preferences indicated by the sculpted pattern, an intermediate value between 0 and 1 should be chosen for λ. By using the revised sensitivity number from Eq. (6.11), the relative importance of elements is adjusted according to the designer's subjective preferences. This enables the designer to steer the final structural topology towards being performance-driven (when λ is close to 0) or preference-driven (when λ is close to 1).

In Step 4, an efficient smoothing technique developed by Li et al. (2022) is applied to transform zigzagged boundaries in the finite element model shown in Fig. 6.9c into a smooth geometry (see Fig. 6.9d). This substantially enhances the aesthetic quality of the optimization result and assists the designer in refining their subjective preferences for the next design exploration cycle.

It is important to note that a single design exploration cycle of the four steps may not yield a satisfactory solution, as the designer often develops new insights and refined preferences during the exploration process. Our human–computer interaction platform allows the designer to modify the solution from the previous design cycle and repeat Steps 1–4, as shown in Fig. 6.11. This iterative approach ensures that the designer's evolving subjective preferences are continually updated until a desired outcome is obtained.

6.5 Performing Interactive Topology Optimization in Virtual Reality

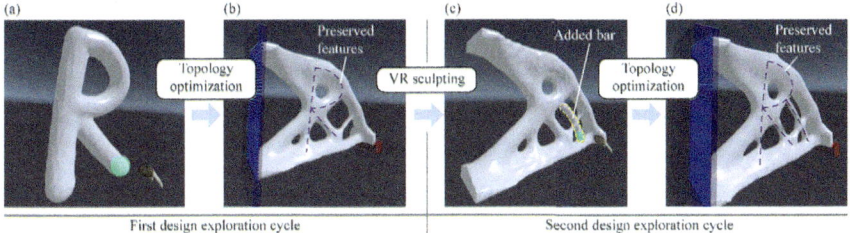

Fig. 6.11 Interactive design exploration cycles: **a** sculpting a preferred pattern using VR; **b** obtaining a solution in the first design exploration cycle; **c** modifying the previous solution through VR sculpting; **d** generating a new solution in the second design exploration cycle (reprinted from Li et al. 2024a, with permission from authors)

We demonstrate the VR-based interactive topology optimization workflow using an example of design exploration for a 3D pavilion, as shown in Fig. 6.12. A cuboidal bounding box of 6 m × 6 m × 4 m, containing a layer of 0.3 m thick non-design domain at the top and a design domain underneath (see Fig. 6.12a), is discretized into 144,000 cubic elements, each with an edge length of 0.1 m. A vertical load totalling 10,000 N is uniformly distributed across the top surface of the non-design domain, while two rectangular areas of fixed supports are assigned to the base of the design domain. BESO topology optimization parameters are $ER = 2\%$, $V^* = 15\%$, and $r_{min} = 0.3$ m. Figure 6.12b shows the optimization result without considering subjective preferences (i.e., $\lambda = 0$).

In this example, our goal is to use VR sculpting to incorporate subjective preferences in the design exploration process while maintaining a high level of structural performance. Figure 6.12c illustrates the starting point and results of the first design exploration cycle, showcasing three innovative structures (Solutions A–C). These designs consider an 'A' shape as the preferred pattern, created through VR sculpting and refined using the CAD software, Rhino. The three structures are optimized using different values of λ (0.04, 0.25, and 0.64). It can be observed that as λ increases, the preferred 'A' shape becomes more pronounced in the resulting designs, but at the cost of lower stiffness (i.e., higher compliance, C). Although Solutions B and C retain key features of the preferred pattern, their compliance values are 7% and 34% higher than that of Solution A, respectively. On balance, Solution B could be an acceptable compromise, but some of its complex geometric details may not meet the designer's expectations.

In the second design exploration cycle, we choose Solution A as the starting point, based two considerations: first, it is the stiffest structure from the previous design cycle, and second, it is less complex than Solutions B and C. By visualizing Solution A from various perspectives in the VR environment, we apply new subjective preferences through VR sculpting. Specifically, we add a horizontal bar to complete the letter 'A' and remove four slender bars near the top, based on our intuitive assessment. During the subsequent topology optimization, we set λ to 0.64—a relatively high value—to ensure that the newly updated geometric details would effectively

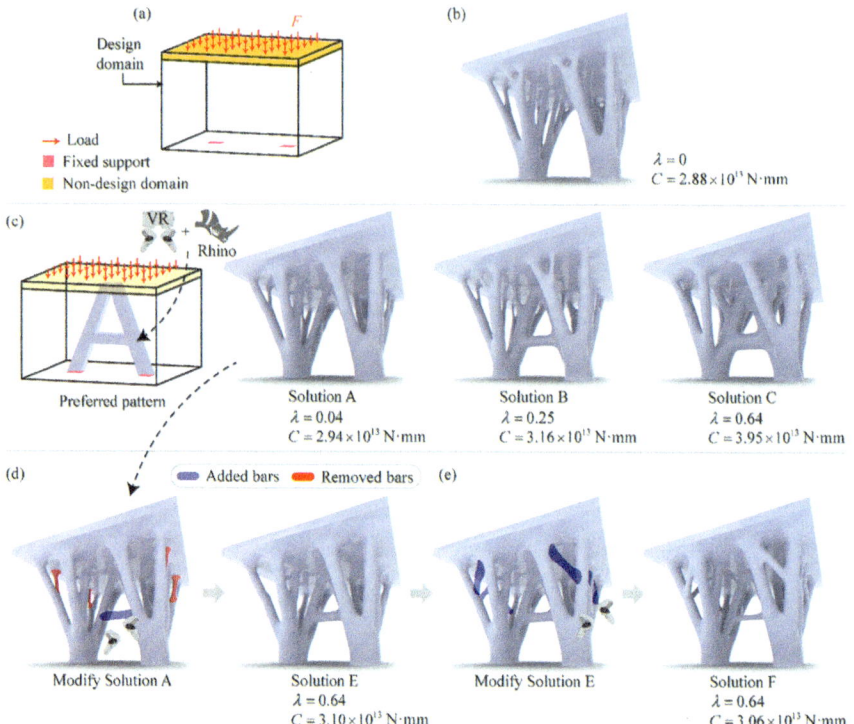

Fig. 6.12 Interactive design exploration using VR for a 3D pavilion: **a** problem setting; **b** reference design obtained from the conventional BESO method; **c** a preferred pattern created by VR sculpting and Rhino, and results of the first design cycle; **d** results of the second design cycle; **e** results of the third design cycle (reprinted from Li et al. 2024a, with permission from authors)

influence the outcome. Indeed, the optimization result (Solution E) keeps all these new modifications. Furthermore, Solution E is stiffer than the most acceptable design from the previous design cycle (Solution B). Both solutions maintain the preferred pattern and have relatively high stiffness. They emerge as strong contenders for the final design.

After a thorough subjective evaluation of geometric details in Solutions B and E in the VR environment, we conclude that Solution B seems overly complex, while solution E appears too plain. Therefore, a third design exploration cycle is conducted (see Fig. 6.12e). In this round, we start with Solution E and modify it by adding eight bars to connect some of the neighbouring bars supporting the roof. The subsequent topology optimization result, Solution F, obtained using $\lambda = 0.64$, not only delivers a visually more appealing design but is also stiffer than both Solutions B and E. Thus, Solution F is selected as the final design for the 3D pavilion.

6.6 Improving Control of Subjective Preferences

In the optimization process described in the previous section, the user can intuitively indicate a level of influence of the preferred pattern by assigning a fixed value to the parameter λ, which is utilized in Eq. (6.11). However, as shown in the pavilion design example, it is challenging to set an 'appropriate' value for λ and the user is unable to accurately control the similarity between the preferred pattern and the optimization result. To address this issue, we introduce an additional constraint on the similarity between the optimized design and the preferred pattern (Li et al. 2025). The optimization problem presented in Sect. 4.2.1 is reformulated as follows

$$\text{Minimize}: C = \mathbf{f}^T \mathbf{u} \tag{6.12a}$$

$$\text{Subject to}: \mathbf{Ku} = \mathbf{f} \tag{6.12b}$$

$$: \sum_{i=1}^{M} V_i x_i = V^* \tag{6.12c}$$

$$: Z = \frac{\sum_{i=1}^{M} z_i x_i}{\sum_{i=1}^{M} z_i} \geq Z^* \tag{6.12d}$$

$$: x_i = x_{\min} \text{ or } 1 \tag{6.12e}$$

$$: z_i = 0 \text{ or } 1 \tag{6.12f}$$

where Z and Z^* represent the current similarity and the user-specified target similarity, respectively. Here, the similarity measure, Z, is defined as the overlapping ratio between the current design and the preferred pattern, as shown in Eq. (6.12d). For each of the M elements in the structural model, if the element is solid, $x_i = 1$; if the element is void, $x_i = x_{\min}$; if the element is part of the preferred pattern, $z_i = 1$; and if the element is outside the preferred pattern, $z_i = 0$. Typically, $x_{\min}$ is set to 0.001 in the 'soft-kill' BESO method (Huang and Xie 2010), but in Eq. (6.12e), $x_{\min}$ can be simply set to 0.

To produce an optimized design that satisfies the above similarity constraint, we replace the fixed parameter λ in Eq. (6.11) with a dynamic parameter $\hat{\lambda}$, which is defined as

$$\hat{\lambda}_{k+1} = \hat{\lambda}_k + \frac{(Z^* - Z)}{Z^*} \tag{6.13}$$

where $\hat{\lambda}_k$ and $\hat{\lambda}_{k+1}$ are the dynamic parameters at the kth and $(k + 1)$th iterations, respectively, with the initial value $\hat{\lambda}_0$ set to 0. Note that $\hat{\lambda}$ ranges from 0 to 1. Therefore, $\hat{\lambda}_{k+1}$ is set to 0 if $\hat{\lambda}_k+(Z^*-Z)/Z^*$ is less than 0, and $\hat{\lambda}_{k+1}$ is set to 1 if $\hat{\lambda}_k+(Z^*-Z)/Z^*$ is greater than 1. By iteratively performing topology optimization and dynamically adjusting the subjective preference weight parameter $\hat{\lambda}$ using Eq. (6.13), we can achieve efficient structural designs that meet the prescribed level of similarity to the preferred pattern, as given in Eq. (6.12d).

It is worth noting that this optimization process may sometimes produce excessively high subjective weights locally within parts of the design domain, leading to the formation of isolated regions. To address this issue, the breadth-first search algorithm is employed to group all interconnected elements at the end of each iteration (Xiong et al. 2023). Then, the largest cluster of elements is preserved while isolated small clusters of elements are removed, thus ensuring the continuity of the structural design.

We use a 3D cantilever example shown in Fig. 6.13 to demonstrate the effectiveness of the new algorithm and the influence of the prescribed target similarity Z^* on the resulting structural topology. The design domain measures 80 mm in length, 20 mm in width, and 50 mm in height, and is discretized into 80,000 cubic elements, each with an edge length of 1 mm. A vertical point load of 1 N is applied downwards at the centre of the free end, while a fixed boundary condition is assigned to the entire rear surface of the cantilever. The material is assumed to be isotropic and linearly elastic, with Young's modulus of 1 MPa and Poisson's ratio of 0.3. The BESO parameters are set as $ER = 3\%$, $V^* = 15\%$, and $r_{\min} = 3$ mm.

Inspired by the shape of a butterfly, we choose the 3D geometry shown in Fig. 6.13a as the preferred pattern. This pattern is initially created through VR sculpting and then refined using the CAD software Rhino.

Figure 6.13b shows the conventional BESO result (Reference 1) which is purely performance-driven and does not consider subjective preferences indicated by the preferred pattern. In contrast, Fig. 6.13c gives an extremely preference-driven design (Reference 2), obtained by the optimization process when Z^* is set to 1. Additionally, we find five different designs by setting Z^* to 0.5, 0.6, 0.7, 0.8, and 0.9, as shown in Fig. 6.13d. The structural performance of each design is evaluated by their relative compliance with respect to Reference 1 (i.e., C/C_1). The evolutionary history of the compliance is displayed in Fig. 6.13e, while the variation of C/C_1 with respect to Z^* and variations of $\hat{\lambda}$ and Z with respect to the iteration step are illustrated in Fig. 6.13f–h, respectively. It is observed that increasing Z^* from 0.5 to 0.9 leads to a noticeable rise in compliance, but with an enhanced geometric similarity to the preferred pattern.

Figure 6.13g, h provides valuable insights into the optimization process and its outcome. During early iterations of the BESO process starting from a fully solid design domain, Z gradually decreases from its initial value of 1. In this phase, $\hat{\lambda}$ remains close to 0; therefore, only structural performance is considered in the topology optimization process. This allows the optimization to quickly approach a shape close to the optimal design. Once Z reaches the target value, Z^*, the optimization process begins to balance two usually conflicting goals of subjective preference

6.6 Improving Control of Subjective Preferences

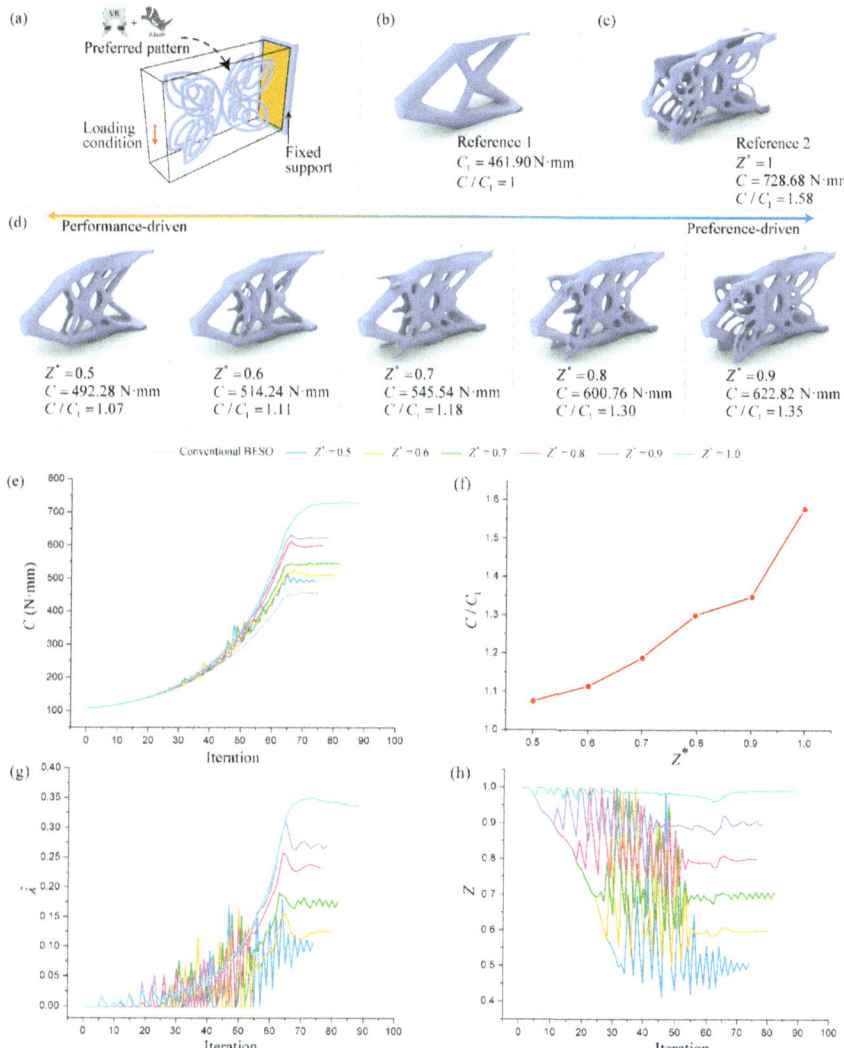

Fig. 6.13 Investigation of the influence of the target similarity Z^* on optimization results of a 3D cantilever: **a** initialization and the preferred pattern created by VR sculpting and Rhino; **b** purely performance-driven design (Reference 1); **c** extremely preference-driven design (Reference 2); **d** five designs obtained by using different Z^* values; **e** evolutionary history of the compliance; **f** variation of C/C_1 with respect to Z^*; **g** variation of $\hat{\lambda}$ with respect to iteration step; **h** variation of Z with respect to iteration step (reprinted from Li et al. 2025, licensed under CC-BY 4.0)

and structural performance, causing $\hat{\lambda}$ to oscillate significantly. As the optimization progresses, these oscillations gradually diminish, and $\hat{\lambda}$ starts to stabilize at a certain level. Consequently, Z also stabilizes and eventually converges to Z^*, ensuring that the final design is not only structurally efficient but also meets the similarity target. This feature is highly desirable, as it enables the designer to effectively and accurately control the geometric similarity between the optimized design and the preferred pattern.

Next, we use a 3D museum design to further demonstrate potential practical applications of the VR-based design exploration strategy. The problem setting of the museum design is inspired by the well-known Qatar National Convention Centre (Qatar Foundation 2024), which was designed by the renowned Japanese architect Arata Isozaki and his collaborators (Cui et al. 2003). In this example, a cuboidal bounding box measuring 200 m in length, 15 m in width, and 22 m in height is chosen, which includes a layer of 1.5 m thick non-design domain on the top and a design domain underneath, as shown in Fig. 6.14a. The box is discretized into 528,000 cubic elements, each with an edge length of 0.5 m. A uniformly distributed vertical load of 10,000 N/m^2 is applied downwards to the top surface of the non-design domain. Fixed supports are assigned to five rectangular areas at the base. The material is assumed to be isotropic and linearly elastic, with Young's modulus of 30 GPa and Poisson's ratio of 0.3. The BESO parameters are set as $ER = 2\%$, $V^* = 12\%$, and $r_{\min} = 1.5$ m.

The preferred pattern in this case consists of 10 arches that connect the five support areas (see Fig. 6.14a). This pattern is initially created through VR sculpting and then refined using the CAD software Rhino. Figure 6.14b shows the conventional BESO result, which does not consider the preferred pattern. This result serves as a reference design to measure the performance loss of other solutions.

Figure 6.14c, d illustrates the outcomes of the first design exploration cycle: Solutions A and B, obtained by setting Z^* to 0.6 and 0.9, respectively. From these results, Solution B may be selected as the preferred design because it retains most of the geometric features of the original preferred pattern and, unlike Solution A, does not contain very slender components. Although Solution A has lower compliance, the slender components in this structure are prone to local buckling, which compromises the safety of the design.

In the second design exploration cycle (see Fig. 6.15a, b), we try to refine the design by modifying some of the details in Solution B based on intuition and subjective preference. First, we use VR sculpting to remove four components (marked in red in Fig. 6.15a) and set the modified geometry as the new preferred pattern. Then, we re-run the optimization using a relatively high Z^* value of 0.95 to ensure that the newly updated geometric features effectively influence the outcome. The optimization result, Solution C shown in Fig. 6.15b, successfully retains all the features of the new preferred pattern. Moreover, Solution C exhibits better structural performance than the preferred design from the previous design cycle (Solution B).

In the third design exploration cycle, we refine the design further by modifying some of the details in Solution C, again based on intuition and subjective preference. This involves adding four straight bars to support the roof (see Fig. 6.15c). Then, the

6.6 Improving Control of Subjective Preferences

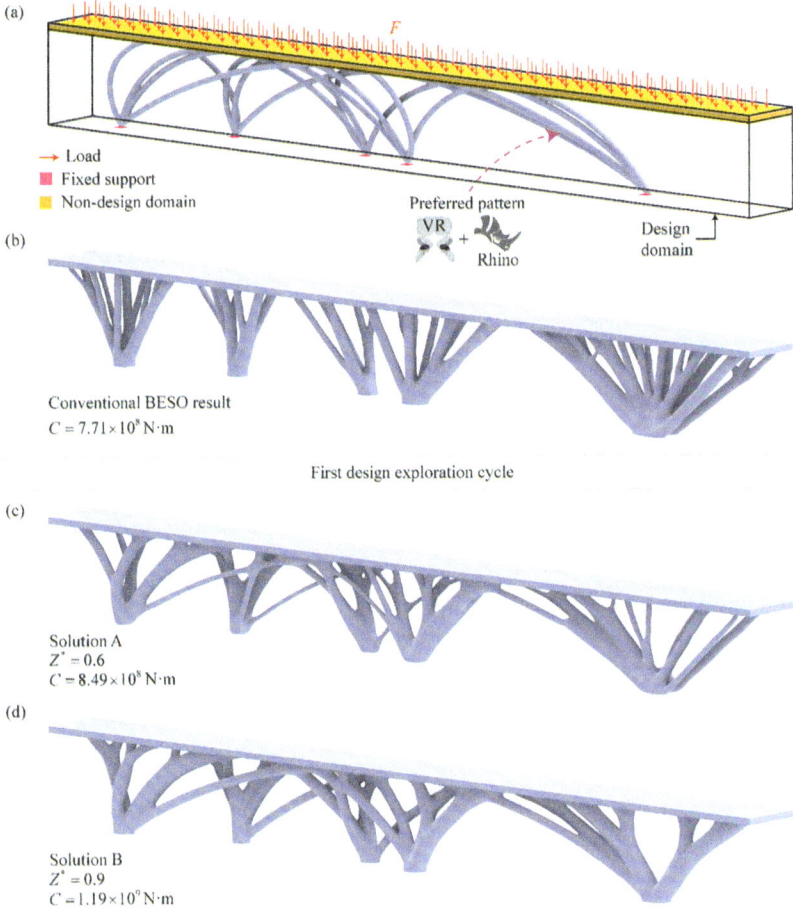

Fig. 6.14 The first design exploration cycle of a museum design: **a** problem setting, and the preferred pattern created by VR sculpting and Rhino; **b** reference design generated by the conventional BESO method; **c** Solution A with $Z^* = 0.6$; **d** Solution B with $Z^* = 0.9$ (reprinted from Li et al. 2025, licensed under CC-BY 4.0)

modified Solution C is taken as the new preferred pattern to guide the next round of topology optimization. The optimized result, Solution D shown in Fig. 6.15d, achieved with $Z^* = 0.8$, presents a visually more appealing design (in the subjective opinion of this designer). In terms of structural performance, the compliance of Solution D is 9.2% and 4.4% lower than those of Solutions B and C, respectively. After weighing both subjective preference and structural performance, Solution D is selected as the final design for the museum. The rendering of this design, shown in Fig. 6.16, appears quite spectacular. This image has been selected as the background for the cover of this book.

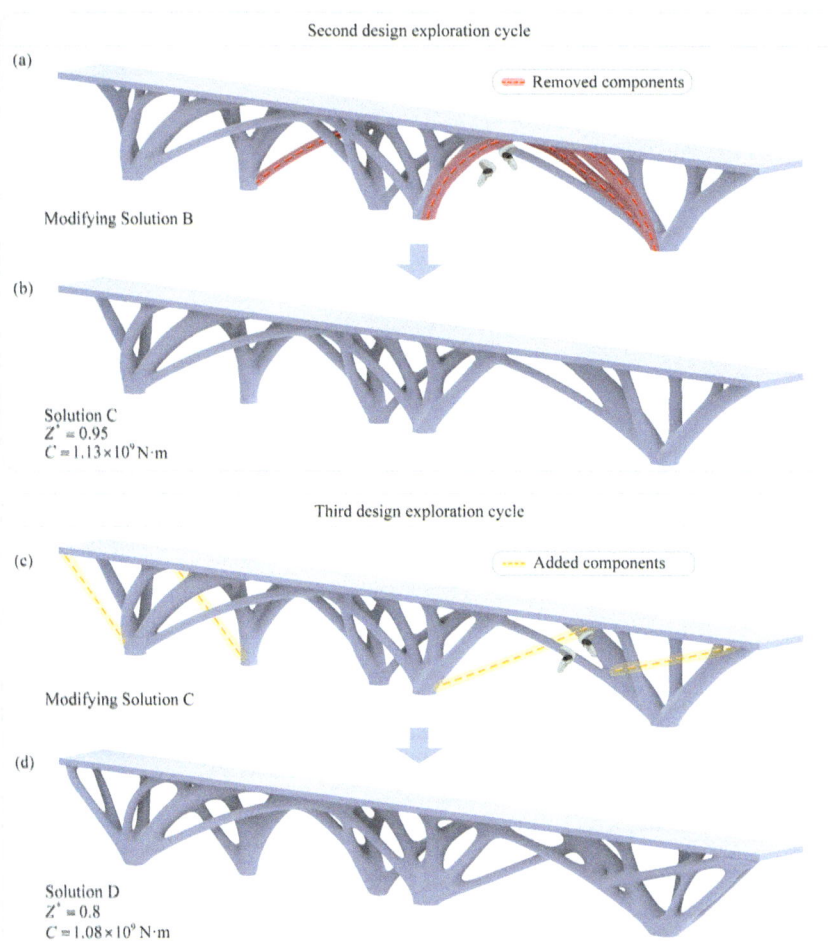

Fig. 6.15 The second and third design exploration cycles of the museum design: **a** modifying Solution B; **b** Solution C with $Z^* = 0.95$; **c** modifying Solution C; **d** Solution D with $Z^* = 0.8$ (reprinted from Li et al. 2025, licensed under CC-BY 4.0)

This example clearly demonstrates the benefits of our iterative design exploration approach based on human–computer collaboration. The designer can influence, or even control, the optimization results by inputting a preferred pattern at the outset and modifying the geometric details of intermediate designs. The results from each design cycle may inspire new ideas, prompting the designer to refine their subjective preferences. Ultimately, the designer may achieve a final solution that not only exhibits high structural performance but also aligns with their artistic vision and satisfies other functional requirements.

Fig. 6.16 Rendering of the museum design (reprinted from Li et al. 2025, licensed under CC-BY 4.0)

6.7 Conclusion

In this chapter, we have presented a design exploration strategy based on human–computer interaction to achieve innovative and efficient structural designs that take subjective preferences into account. This strategy can effectively harness the complementary strengths of human creativity and computational power. We have developed two techniques that enable the designer to influence or control the design outcomes: one involves drawing or importing a preferred pattern, while the other assigns scores to intermediate designs. These subjective preferences are incorporated in a modified topology optimization algorithm to steer the evolution of the structural design towards a preferred form.

For the design of complex 3D structures, we have developed a VR-based topology optimization method and demonstrated how the VR environment offers an interactive, intuitive, and immersive platform for the designer to visualize and edit 3D geometries. Additionally, we have developed an algorithm that enables the designer to effectively and accurately control the geometric similarity between the optimized design and the preferred pattern.

It is worth noting that the VR-based topology optimization can be further extended to a mixed reality (MR) setting, allowing virtual models to interact seamlessly with the physical environment in real time (Jahn et al. 2018; Kraus et al. 2022). The MR extension can provide the designer with more interactive and practical possibilities by blending the digital and physical worlds (Li 2024).

References

Chen, C.-W., Hu, M.-C., Chu, W.-T. and Chen, J.-C. (2021) A real-time sculpting and terrain generation system for interactive content creation. *IEEE Access* **9**, 114914–114928.

Cui, C., Ohmori, H. and Sasaki, M. (2003) Computational morphogenesis of 3D structures by extended ESO method. *J. Int. Assoc. Shell Spat. Struct.* **44**, 51–61.

Holland, J. H. (1975) *Adaptation in Natural and Artificial Systems: An Introductory Analysis with Applications to Biology, Control, and Artificial Intelligence*. Ann Arbor: University of Michigan Press.

Huang, X. and Xie, Y. M. (2010) *Evolutionary Topology Optimization of Continuum Structures: Methods and Applications*. Chichester: John Wiley & Sons.

Jahn G., Newnham, C., van den Berg, N. and Beanland, M. (2018) Making in mixed reality: Holographic design, fabrication, assembly and analysis of woven steel structures. *Proc. 38th Annul. Conf. ACADIA*, Mexico City, 18–20 October 2018, 88–97.

Kazhdan, M., Bolitho, M. and Hoppe, H. (2006) Poisson surface reconstruction. *Proc. 4th Eurographics Symp. Geometry Process.*, Cagliari, 26–28 June 2006, 61–70.

Kirkland, E. J. (2010) Bilinear interpolation. In *Advanced Computing in Electron Microscopy*. Boston: Springer, 261–263.

Kraus, M. A., Čustović, I. and Kaufmann, W. (2022) Mixed reality applications for teaching structural design. *Proc. Struct. Congress*, Atlanta, 20–23 April 2022, 283–295.

Li, Z. (2024) *Interactive Structural Topology Optimisation Considering Subjective Preferences*. PhD thesis, RMIT University, Australia.

Li, Z., Lee, T.-U., Yao, Y. and Xie, Y. M. (2022) Smoothing topology optimization results using pre-built lookup tables. *Adv. Eng. Softw.* **173**, 103204.

Li, Z., Lee, T.-U. and Xie, Y. M. (2023a) Interactive structural topology optimization with subjective scoring and drawing systems. *Comput. Aided Des.* **160**, 103532.

Li, Z., Lee, T.-U. and Xie, Y. M. (2023b) iBESO: An interactive 2D topology optimization software. https://albertlidesign.gitbook.io/ibeso. Accessed 8 December 2024.

Li, Z., Lee, T.-U. and Xie, Y. M. (2023c) Topology optimisation considering subjective preferences: current progress and challenges. *Proc. IASS Annu. Symp.*, Melbourne, 10–14 July 2023.

Li, Z., Lee, T.-U. and Xie, Y. M. (2024a) Exploring and optimizing innovative structures in virtual reality. *Proc. IASS Annu. Symp.*, Zurich, 26–30 August 2024.

Li, Z., Lee, T.-U. and Xie, Y. M. (2024b) VR-BESO: A VR-based structural design tool. https://albertlidesign.gitbook.io/vr-beso/. Accessed 8 December 2024.

Li, Z., Lee, T.-U. and Xie, Y. M. (2025) Interactive 3D structural design in virtual reality using preference-based topology optimization. *Comput. Aided Des.* **180**, 103826.

Ma, J., Li, Z., Zhao, Z.-L. and Xie, Y. M. (2021) Creating novel furniture through topology optimization and advanced manufacturing. *Rapid Prototyp. J.* **27**, 1749–1758.

Martin H. (2018) The story behind Frank Gehry's iconic Wiggle design. *Archit. Dig.* https://www.architecturaldigest.com/story/the-story-behind-frank-gehrys-iconic-wiggle-design. Accessed 8 December 2024.

Meta (2024) Meta Quest 3: Ultimate power meets premium comfort. https://www.meta.com/au/quest/quest-3/. Accessed 8 December 2024.

Qatar Foundation (2024) Qatar National Convention Centre. https://www.qf.org.qa/community/qatar-national-convention-centre. Accessed 8 December 2024.

Robert McNeel & Associates (2024) Rhino 8. https://www.rhino3d.com/. Accessed 8 December 2024.

Wang, B., Zhou, Y., Zhou, Y, Xu, S. and Niu, B. (2018) Diverse competitive design for topology optimization. *Struct. Multidisc. Optim.* **57**, 891–902.

Wolfartsberger, J. (2019) Analyzing the potential of Virtual Reality for engineering design review. *Autom. Constr.* **104**, 27–37.

Xiong, Y., Zhao, Z.-L., Lu, H., Shen, W. and Xie, Y. M. (2023). Parallel BESO framework for solving high-resolution topology optimisation problems. *Adv. Eng. Softw.* **176**, 03389.

Open Access This chapter is licensed under the terms of the Creative Commons Attribution 4.0 International License (http://creativecommons.org/licenses/by/4.0/), which permits use, sharing, adaptation, distribution and reproduction in any medium or format, as long as you give appropriate credit to the original author(s) and the source, provide a link to the Creative Commons license and indicate if changes were made.

The images or other third party material in this chapter are included in the chapter's Creative Commons license, unless indicated otherwise in a credit line to the material. If material is not included in the chapter's Creative Commons license and your intended use is not permitted by statutory regulation or exceeds the permitted use, you will need to obtain permission directly from the copyright holder.

Chapter 7
Practical Applications

This chapter presents a series of practical examples to illustrate potential real-world applications of the generalized topology optimization approach discussed in this book. In particular, we focus on two distinct cases—one primarily driven by structural performance and the other mainly concerned with aesthetics. The first case involves the design of a long-span steel–concrete composite bridge, while the second explores the creation of an innovative chair. These examples demonstrate that the concepts and techniques introduced in earlier chapters of this book can be employed to create designs that are both practically viable and aesthetically appealing.

7.1 Introduction

Over the past 10 years, I have been fortunate to collaborate with many talented practising architects and engineers, applying our topology optimization techniques to various real-world projects. Figure 7.1 shows a large-scale architectural complex named 'Xiong'an Wings'. In collaboration with Tongji Architectural Design (Group) Co. Ltd., we optimized the core structural system using a multi-material bi-directional evolutionary structural optimization (BESO) method (Li et al. 2023). Construction of this project was completed in 2024.

Figure 7.2 illustrates a 470 m span network arch bridge recently designed for the Kui Zhou Yangtze River Bridge. In collaboration with T.Y. Lin International Engineering Consulting (China) Co. Ltd., we applied the multi-material BESO method to the topology optimization of the wind bracings that connect the two main arches, and to the weight reduction of four spherical bearings that support the arches (Lai et al. 2023). Construction of this bridge is planned to commence in the near future.

My active involvement in real-world projects and frequent dialogue with practising architects and engineers have inspired much of the thinking and many of the techniques presented in this book. In particular, I have been repeatedly reminded that

© The Author(s) 2025
Y. M. Xie, *Generalized Topology Optimization for Structural Design*,
https://doi.org/10.1007/978-981-96-4524-4_7

Fig. 7.1 Rendering of a five-storey architectural complex named 'Xiong'an Wings', with its core structural system optimized using the BESO method. Construction completed in 2024 (reprinted from Li et al. 2023, licensed under CC-BY 4.0)

Fig. 7.2 Rendering of the Kui Zhou Yangtze River Bridge, with its wind bracings and arch bearings optimized using the BESO method (reprinted from Lai et al. 2023, with permission from Elsevier)

a topologically optimized design focused solely on structural performance is often unsuitable for practical implementation and often fails to meet the aesthetic preferences of the designer or client. As a result, I strongly believe that when working on practical projects, our mindset should shift from seeking the 'globally optimal' solution with the best structural performance to flexibly exploring diverse and competitive designs.

In the following sections, we use two examples as case studies to demonstrate that the generalized topology optimization approach introduced in this book can be employed to produce designs that are not only practically viable but also aesthetically appealing. Section 7.2 presents the design of a long-span steel–concrete composite bridge, while Sect. 7.3 discusses the design of an innovative chair.

7.2 Design of a Long-Span Steel–Concrete Composite Bridge

Figure 7.3 shows the final design of a steel–concrete composite bridge with a main span of 350 m and two side spans of 195 m each (Li et al. 2022). The width of the deck is 12 m. Since the bridge's length-to-width ratio is relatively large and the dominant loads act primarily in the vertical plane, the structural model can be simplified into a 2D problem when performing the conceptual design for the bridge using topology optimization.

The design outcome shown in Fig. 7.3 is the result of close collaboration and multiple revisions between researchers in topology optimization and experienced engineers from T. Y. Lin International Engineering Consulting, a leading bridge design firm. In this project, we employ a multi-material BESO method (Li and Xie 2021a, b) and utilize concepts from Chap. 2 on creating diverse and competitive designs, as well as from Chap. 3 on redefining the design domain. Specifically, we strategically explore a variety of design domains and non-design domains, and progressively refine the solutions by considering both structural rationality and aesthetic quality.

The first set of results shown in Fig. 7.4 is obtained by assuming the design domain to be below the bridge deck. Due to the overall symmetry of the bridge, only

Fig. 7.3 Rendering of a long-span steel–concrete composite bridge, optimized using a multi-material BESO method (reprinted from Li et al. 2022, with permission from Elsevier)

the left half is considered in the finite element models for topology optimization (see Fig. 7.4a, c, e, g). The optimized designs are illustrated for the full bridge in Fig. 7.4b, d, f, h, with red and blue colours representing tensile (steel) and compressive (concrete) areas, respectively. It is important to note that in Fig. 7.4c, e, g, symmetric regions (marked in pink) are specified, forcing the resulting topologies to be symmetric about the main pier, unlike the solution in Fig. 7.4b. This constraint on geometric symmetry about the pier is a critical requirement of the construction process, which is explained later.

After discussing the preliminary results shown in Fig. 7.4 with the lead bridge engineer on the project team, we realize that these designs would be difficult to implement in practice due to the discontinuity of the steel cables. This problem is largely caused by the very limited design space below the deck. To address this issue, in the next cycle of design exploration, we allow the design domain to expand above the deck, in various shapes ranging from a rectangle to a cambered area, as shown in Fig. 7.5a, c, e, g. The optimization results are given in Fig. 7.5b, d, f, h.

It is evident that the new topologies are substantially different from those in Fig. 7.4. Most significantly, we find a continuous steel cable going across the bridge,

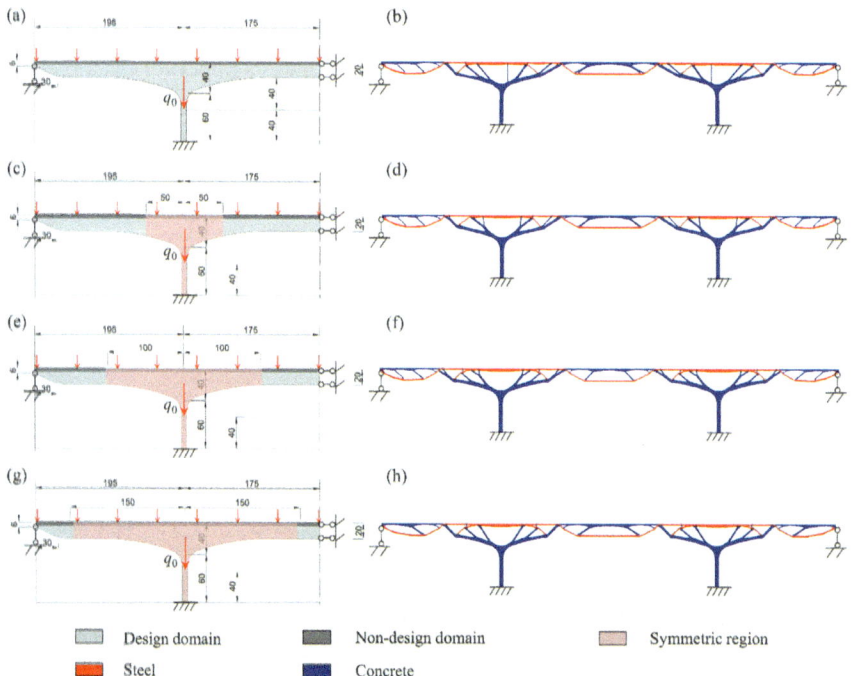

Fig. 7.4 Solutions with the design domain set below the bridge deck (unit: m): **a, b** without a symmetric region; **c, d** with a symmetric region imposed, 2 × 50 m in width; **e, f** with a symmetric region imposed, 2 × 100 m in width; **g, h** with a symmetric region imposed, 2 × 150 m in width (reprinted from Li et al. 2022, with permission from Elsevier)

7.2 Design of a Long-Span Steel–Concrete Composite Bridge

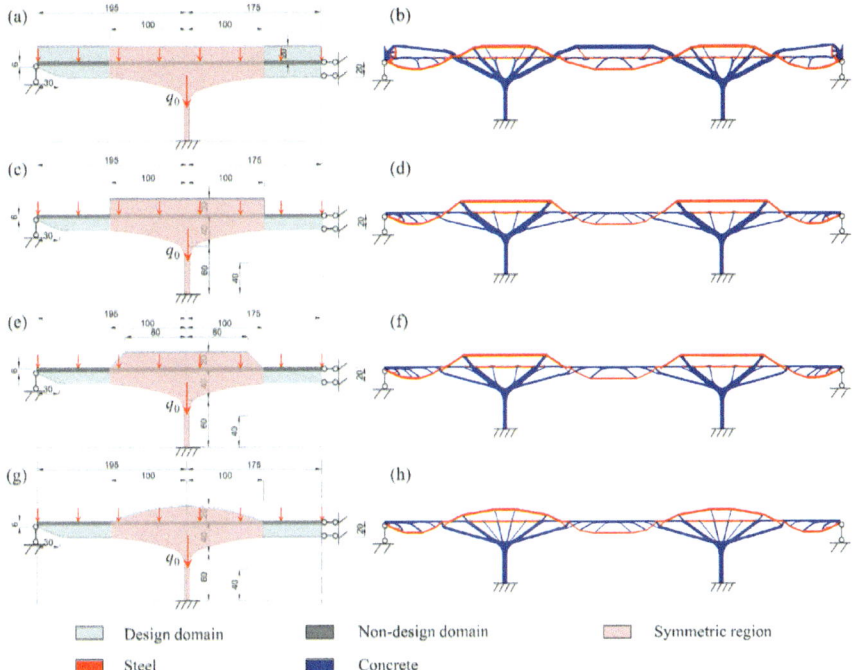

Fig. 7.5 Solutions corresponding to various expansions of the design domain above the deck (unit: m): **a, b** a rectangle with a width of 370 m; **c, d** a rectangle with a width of 200 m; **e, f** a trapezoid; **g, h** a cambered area (reprinted from Li et al. 2022, with permission from Elsevier)

effectively carrying most of the tensile forces. It is also noted that, by having a cambered area above the deck in the expanded design domain, we achieve a smooth and elegant curve for the main steel cable, as shown in Fig. 7.5h.

In the third cycle of design exploration, we retain the cambered area in the design domain and add various geometric features to the non-design domain, such as an arch on each side of the pier (see Fig. 7.6c), a vertical middle column (see Fig. 7.6e), and both the arches and the column (see Fig. 7.6g). These geometric features are explored based on aesthetic considerations of the project team. Ultimately, the topology shown in Fig. 7.6d is selected as the preferred design. After making minor adjustments, we arrive at the final conceptual design for the bridge, as illustrated in Fig. 7.3.

It is noted that in Figs. 7.4, 7.5 and 7.6, a symmetric constraint is imposed in the pink area of the design domain in each of the optimization models, except for the one in Fig. 7.4a. This geometric symmetry about each main pier is dictated by the construction process of the bridge (see Figs. 7.7 and 7.8). Such a high-pier and long-span bridge is typically built using a balanced cantilever segmental construction method with form travellers (see Fig. 7.7). To balance the heavy loads from the structural members' self-weight and the form travellers, it is essential for each half of the bridge to maintain a symmetric shape during construction. Additionally, temporary

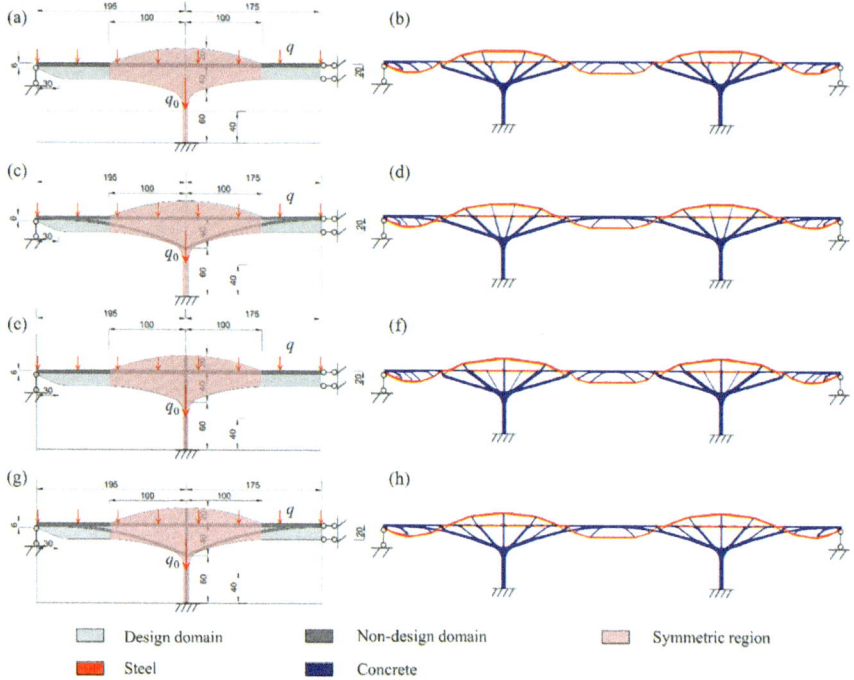

Fig. 7.6 Solutions corresponding to different non-design domains (unit: m): **a, b** no additional non-design domain other than the deck; **c, d** with an arch added on each side of the pier to the non-design domain; **e, f** with a vertical middle column added to the non-design domain; **g, h** with both the arches and the vertical middle column added to the non-design domain (reprinted from Li et al. 2022, with permission from Elsevier)

cables are used to help the main girder and lower cord arch resist excessive bending moments before the two halves of the bridge are joined together in later stages of the construction process (see Fig. 7.8).

It should be noted that when topology optimization is employed for the conceptual design of a large-scale complex structure, only the predominant loads or load cases are considered duing the form-finding process. Subsequently, a more detailed analysis accounting for all the static and dynamic load conditions as required by local or international design codes must be performed before the design can be implemented. If the design does not satisfy all the requirements, modifications need to be made to the structure, followed by a thorough reanalysis.

For the final bridge design shown in Fig. 7.3, the detailed analysis we have conducted includes considerations of traffic loads, wind loads, temperature effects, dynamic response, and buckling resistance (Li et al., 2022).

7.3 Design of an Innovative Chair

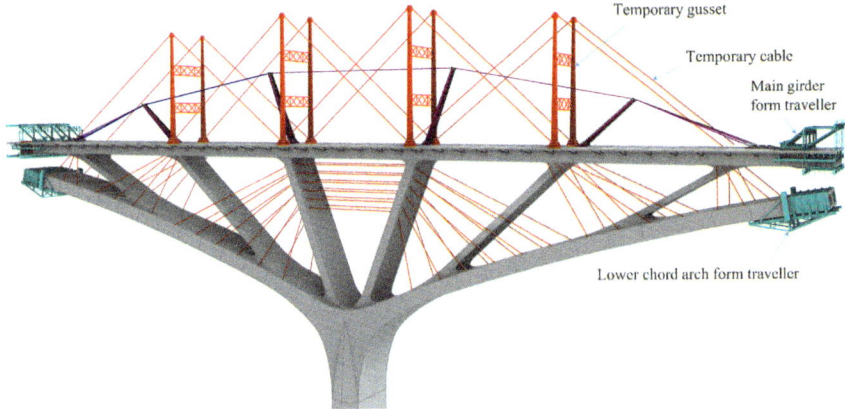

Fig. 7.7 Schematic diagram of the construction process for one half of the bridge (reprinted from Li et al. 2022, with permission from Elsevier)

7.3 Design of an Innovative Chair

While bridge design is primarily driven by structural performance, furniture design is often dictated by subjective preferences. The style and aesthetics of a chair are typically more important than its structural function of supporting a person's weight. In this context, we demonstrate that the generalized topology optimization approach can be used to create innovative chair designs that strike a fine balance between structural requirements and aesthetic preferences.

Figure 7.9 illustrates a topologically optimized, 3D-printed chair created by my team (Ma et al. 2021). The lead designer, Zhi Li, had a professional background in fine arts before pursuing a Ph.D. in engineering under my supervision. By integrating aesthetic considerations and engineering principles, we were able to produce this chair that has won multiple international awards.

The initial design for the chair shown in Fig. 7.10 is inspired by an ancient Chinese wine vessel, *Jue* (爵) (Childs-Johnson 1987). This model is created using the computer-aided design (CAD) software Rhino (Robert McNeel & Associates 2024). We first generate a low-resolution CAD model (see Fig. 7.10a) using the hierarchical subdivision surface modelling technique (Bhooshan and El Sayed 2011). Then, we apply the Catmull–Clark subdivision scheme (Stam 1998) to smooth the model and create a high-resolution CAD model (see Fig. 7.10b). The quadrilateral meshes in these models enable easy and fast geometric modifications.

As demonstrated in Chap. 3, the settings of the design domain and non-design domain significantly influence the design outcome. In this example, much of the design domain is predetermined by the choice of the initial design, which is based on the designer's subjective preferences. Additionally, the edges of the initial design are designated as the non-design domain, ensuring that the optimization result preserves the outlines of the preferred shape. For boundary and load conditions, fixed supports

Fig. 7.8 Construction sequence of the bridge (reprinted from Li et al. 2022, with permission from Elsevier)

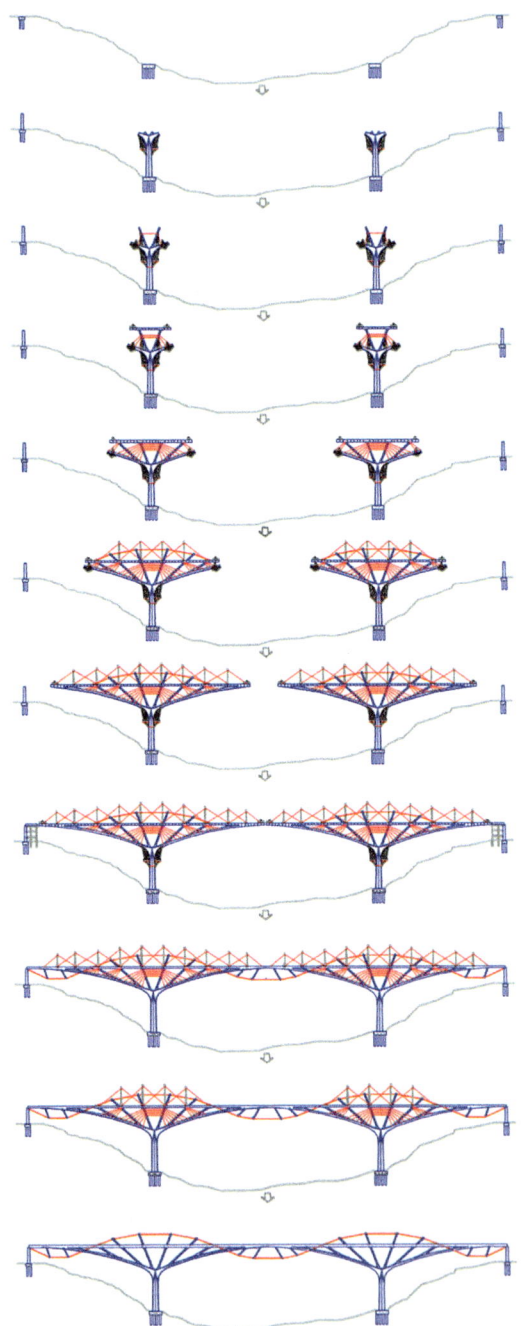

7.3 Design of an Innovative Chair

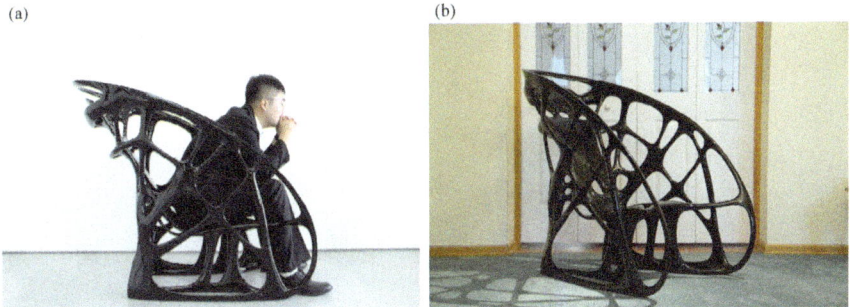

Fig. 7.9 Photos of a topologically optimized, 3D-printed chair (Photo credits: **a** Zhi Li; **b** Shuiqing Zhang. Printed with permission)

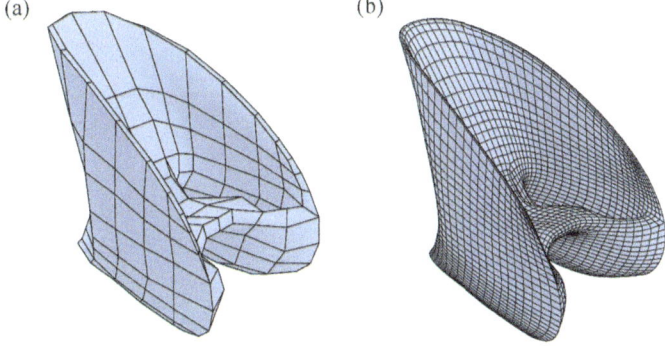

Fig. 7.10 Initial design of the chair: **a** low-resolution CAD model; **b** high-resolution CAD model (reprinted from Ma et al. 2021, with permission from Emerald)

are assigned to the base of the model, and three load cases are considered to simulate different scenarios of a person sitting or lying in the chair (Zhu, 2013).

After discretizing the high-resolution CAD model in Fig. 7.10b into a fine mesh of tetrahedral elements, we perform topology optimization using the Ameba software (Zhou et al. 2018; XIE Technologies 2024), which is a design tool based on the BESO method (Huang and Xie 2010). Figure 7.11 illustrates evolutionary histories of the structural compliance and volume fraction, with the target volume fraction set to 20%.

The topology optimization result, shown in Fig. 7.12a, contains many irregular tetrahedral elements and a serrated surface. This may lead to non-manifold edges and vertices, making the design unsuitable for direct manufacturing. Therefore, post-processing is necessary to achieve a smooth design that meets the surface quality requirements of a practical product. First, we generate a voxel-based signed distance field from the tetrahedral elements (Osher and Fedkiw 2003). Then, we apply an isosurface extraction method called 'marching cubes' for surface reconstruction to obtain a manifold model (see Fig. 7.12b) (Lorensen and Cline 1987).

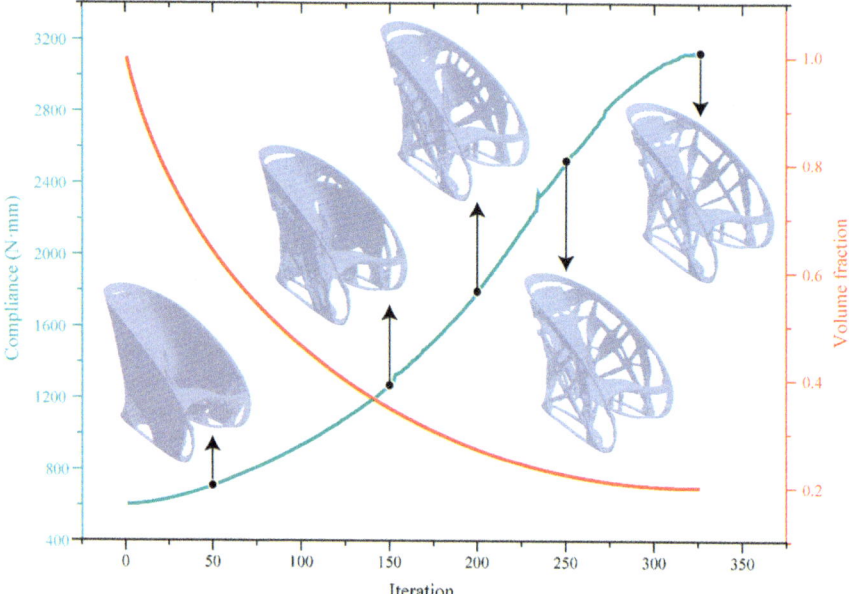

Fig. 7.11 Evolutionary histories of structural compliance and volume fraction (reprinted from Ma et al. 2021, with permission from Emerald)

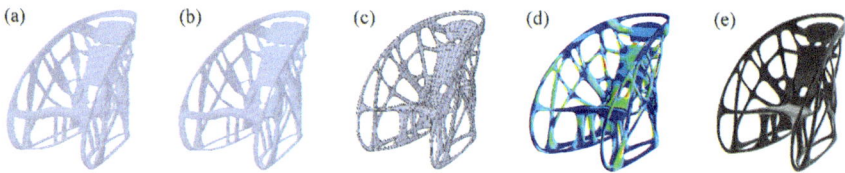

Fig. 7.12 Post-processing and re-evaluation of topology optimization result: **a** optimization result; **b** isosurface extraction; **c** quad remeshing; **d** finite element analysis; **e** rendering of the final design (reprinted from Ma et al. 2021, with permission from Emerald)

However, the surface reconstruction described above may result in numerous irregular triangular facets and a disorganized vertex distribution, which are not ideal for making geometrical modifications to meet aesthetic or functional requirements. To address this issue, we use the QuadRemesh function in Rhino to create a CAD model with a quadrilateral mesh (see Fig. 7.12c). This model can then be easily modified in common CAD software tools such as Rhino. After these design adjustments, finite element analysis is performed on a much finer mesh to re-evaluate the structural performance (see Fig. 7.12d). Ideally, an 'optimal' chair design should balance aesthetics, structural performance, and functional requirements. Figure 7.12e shows the rendering of our final chair design.

It is worth noting that this chair was designed before we developed the human–computer interaction tools presented in Chap. 6. These tools would have been highly beneficial for improving both the design process and the outcome, in two ways. First, rather than specifying a non-design domain that cannot be altered during topology optimization, we could have drawn a preferred pattern and allowed it to evolve into a shape that incorporates both structural performance and aesthetic preferences. The drawn pattern would act as an 'amenable non-design domain', providing greater flexibility in the optimization process. Second, post-processing the topology optimization result, as shown in Fig. 7.12, might inadvertently cause a significant loss in structural performance when details of the optimized design are modified intuitively. With the interactive design approach introduced in Chap. 6, the modified design would undergo another round of topology optimization, ensuring that the structural performance is maintained.

Next, we show in Fig. 7.13 the process of fabricating the chair through additive manufacturing, commonly known as '3D printing'. The chair's dimensions are 1310 mm × 796 mm × 1134 mm. Due to its large size, printing the chair in a single piece would be challenging and time-consuming. Therefore, the chair model is divided into six parts, which are printed separately but simultaneously. Dowel and glue are used to assemble these parts. The fused filament fabrication (FFF) technique is employed because it is relatively inexpensive and suitable for large-scale printing. However, the layer-by-layer printing process of FFF results in a staircase effect on curved or inclined surfaces. To enhance the surface quality of the chair, sanding, polishing, and painting are performed after the parts are printed and assembled. The finished chair exhibits a smooth, gleaming surface and exudes striking elegance (see Fig. 7.9).

7.4 Conclusion

The practical examples presented in this chapter further highlight the immense potential of applying topology optimization to real-world projects. We have repeatedly demonstrated that to create innovative, efficient, and elegant structures using topology optimization, designers should explore diverse and competitive solutions, strategically redefine the design domain, and modify intermediate designs based on subjective preferences or functional requirements—then re-run topology optimization to maintain the structural performance. These are some of the key concepts I have advocated in this book. While some of these concepts may seem trivial, I have found them very useful when working on real-world projects in collaboration with practising architects and engineers. I am confident that the concepts and techniques presented in this book will find many more practical applications in the years to come.

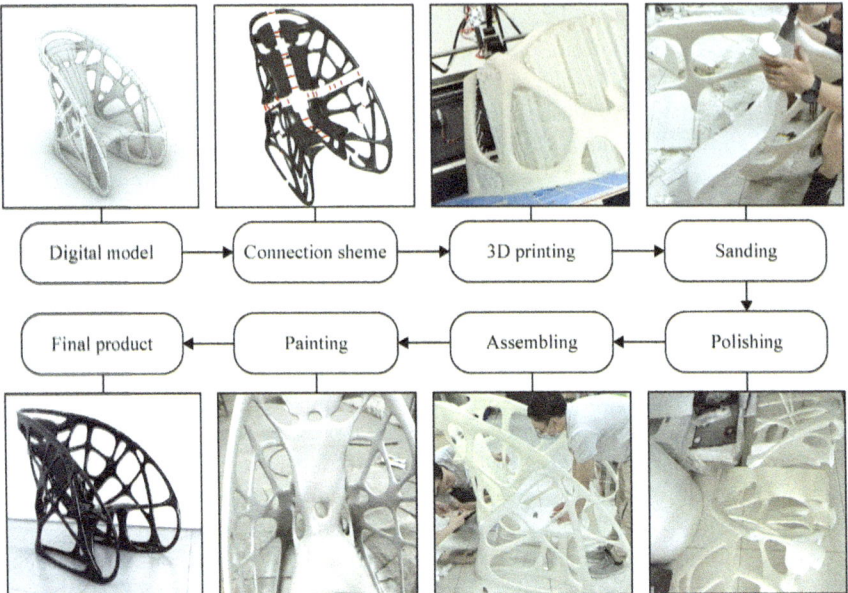

Fig. 7.13 Manufacturing workflow of the chair (reprinted from Ma et al. 2021, with permission from Emerald)

References

Bhooshan, S. and El Sayed, M. (2011) Use of sub-division surfaces in architectural form-finding and procedural modelling. *SimAUD 2011: Proc. 2011 Symp. Simul. Achitect. Urban Des.*, Boston, 3–7 April 2011, 60–67.

Childs-Johnson, E. (1987) The jue and its ceremonial use in the ancestor cult of China. *Artib. As.* **48**, 171–196.

Huang, X. and Xie, Y. M. (2010) *Evolutionary Topology Optimization of Continuum Structures: Methods and Applications.* Chichester: John Wiley & Sons.

Lai, Y., Li, Y., Huang, M., Zhao, L., Chen, J. and Xie, Y. M. (2023) Conceptual design of long span steel–UHPC composite network arch bridge. *Eng. Struct.* **277**, 115434.

Li, Y., Ding, J., Zhang, Z., Zhou, X., Makvandi, M., Yuan, P. F. and Xie, Y. M. (2023) Practical application of multi-material topology optimization to performance-based architectural design of an iconic building. *Compos. Struct.* **325**, 117603.

Li, Y., Lai, Y., Lu, G., Yan, F., Wei, P. and Xie, Y. M. (2022) Innovative design of long-span steel–concrete composite bridge using multi-material topology optimization. *Eng. Struct.* **269**, 114838.

Li, Y. and Xie, Y. M. (2021a) Evolutionary topology optimization for structures made of multiple materials with different properties in tension and compression. *Compos. Struct.* **259**, 113497.

Li, Y. and Xie, Y. M. (2021b) Evolutionary topology optimization of spatial steel–concrete structures. *J. Int. Assoc. Shell Spat. Struct.* **62**, 102–112.

Lorensen, W. E. and Cline, H. E. (1987) Marching cubes: A high resolution 3D surface construction algorithm. *ACM SIGGRAPH Comput. Graph.* **21**, 163–169.

Ma, J., Li, Z., Zhao, Z.-L. and Xie, Y. M. (2021) Creating novel furniture through topology optimization and advanced manufacturing. *Rapid Prototyp. J.* **27**, 1749–1758.

References

Osher, S. and Fedkiw, R. (2003) *Level Set Methods and Dynamic Implicit Surfaces.* New York: Springer.

Robert McNeel & Associates (2024) Rhino 8. https://www.rhino3d.com/. Accessed 8 December 2024.

Stam, J. (1998) Exact evaluation of Catmull-Clark subdivision surfaces at arbitrary parameter values. *SIGGRAPH 1998: Proc. 25th Annu. Conf. Comput. Graph. Interact. Tech.*, Orlando, 19–24 July 1998, 395–404.

XIE Technologies (2024) Ameba: Topology optimization software based on BESO. https://ameba.xieym.com. Accessed 8 December 2024.

Zhou, Q., Shen, W., Wang, J., Zhou, Y. Y. and Xie, Y. M. (2018) Ameba: A new topology optimization tool for architectural design. *Proc. IASS Annu. Symp.*, Boston, 16–20 July 2018.

Zhu, H. (2013) *Modeling of Pressure Distribution of Human Body Load on an Office Chair Seat.* Master thesis, Blekinge Institute of Technology, Sweden.

Open Access This chapter is licensed under the terms of the Creative Commons Attribution 4.0 International License (http://creativecommons.org/licenses/by/4.0/), which permits use, sharing, adaptation, distribution and reproduction in any medium or format, as long as you give appropriate credit to the original author(s) and the source, provide a link to the Creative Commons license and indicate if changes were made.

The images or other third party material in this chapter are included in the chapter's Creative Commons license, unless indicated otherwise in a credit line to the material. If material is not included in the chapter's Creative Commons license and your intended use is not permitted by statutory regulation or exceeds the permitted use, you will need to obtain permission directly from the copyright holder.

Author Index

A
Aage, N., 15, 42, 46
Aida, T., 54
Alexa, M., 87
Allaire, G., 69
Alonso Gordoa, C., 53
Alù, A., 54
Amir, O., 59, 60
Andreassen, E., 15
Ansola Loyola, R., 53
Associates, 119, 139

B
Baker, W. F., 1, 4
Bao, D. W., 9, 10, 12, 32, 44, 77
Baran, I., 15
Beanland, M., 129
Beck, A. T., 53
Beghini, A., 1, 4
Beghini, L. L., 1, 4
Bendsøe, M. P., 1–4, 63
Bhooshan, S., 139
Bi, M., 77
Birker, T., 64
Bolitho, M., 119
Bonner, J., 42
Buhl, T., 61, 70, 74

C
Cai, K., 7, 8
Cai, Q., 7, 34
Chen, C.-W., 119
Cheng, G., 32
Chen, J.-C., 119, 133
Chen, S., 31

Chen, Y., 96
Childs-Johnson, E., 139
Christiansen, A. N., 46
Chu, W.-T., 119
Clausen, A., 42
Cline, H. E., 141
Cohen-Or, D., 87
Coulais, C., 54
Cross, R., 3, 81
Cui, C., 4, 19, 126
Cui, T., 9, 19
Čustović, I., 129

D
da Silva, G. A., 53
Ding, J., 133

E
El Sayed, M., 139

F
Fan, Z., 46
Fedkiw, R., 141
Feng, R., 8, 34
Feng, X. Q., 38, 48–52
Furuta, K., 40

G
Gao, L., 46
Garaigordobil Jiménez, A., 53
Goulart, P. J., 96
Guo, X., 31, 32
Guo, Y., 31

H
Han, H., 31
Hattel, J. H., 15
He, L., 8, 34
He, Y., 7, 8, 18, 21, 32, 33, 39, 40
Holland, J. H., 117
Hoppe, H., 119
Huang, M., 133
Huang, W., 7, 39
Huang, X., 2, 8, 22, 40, 42, 65, 69, 98, 123, 141
Hui, D., 31
Hu, J., 69
Hu, M.-C., 119

I
Imbalzano, G., 31
Ishida, Y., 54
Izui, K., 40

J
Jahn G., 129
Jang, G. W., 59
Jeong, S., 82
Jihong, Z., 60
Jolivet, P., 40
Jouve, F., 69

K
Kanebako Y., 2
Kashani, A., 31
Katz, N., 1, 4
Kaufmann, W., 129
Kawabata, M., 2
Kawaguchi, K., 2
Kawaguchi, M., 2
Kazhdan, M., 119
Kikuchi, N., 1
Kim, C. W., 82
Kim, Y. Y., 59
Kirby, J., 8
Kirkland, E. J., 111
Kitamura, M., 82

Kogiso, N., 82
Kondoh, T., 40
Kraft D., 96
Kraus, M. A., 129
Kumar, T., 63

L
Lai, Y., 133, 135
Lazarov B. S., 14
Lee, T.-U., 45, 61, 62, 88, 97, 101, 108, 109, 111, 114, 115, 118, 119, 121, 123, 125, 127–129
Liang, Y., 32
Lian, H., 46
Li, H., 40, 46
Lin, X., 7
Lipman, Y., 87
Liu, Y., 31
Liu, Z., 31
Li, Y., 38, 133, 135
Li, Z., 4, 54, 108, 109, 111, 114, 115, 118, 119, 121, 123, 125, 127–129, 139
Lorensen, W. E., 141
Lu, G., 135
Lu, H., 8, 97, 101, 124

M
Ma, J., 4, 8, 34, 115, 139
Makvandi, M., 133
Martin H., 115
Matsubara, S., 54
Meng, X., 42–44, 51, 59
Meta, 118
Morales-Beltran, M., 4
Mostafavi, S., 4

N
Nemirovskii A., 96
Nervi, P. L., 9
Nesterov Y., 96
Newnham, C., 129
Ngo, T. D., 31
Nguyen, K. T. Q., 31
Nii, S., 82
Nishiwaki, S., 40
Nishizawa, R., 43
Niu, B., 110

O
Ohmori, H., 4, 19, 126

Author Index

Ohsaki, M., 2
Okumura, D., 54
Osher, S., 141

P
Paulino, G. H., 1, 4
Peng, X., 10, 12, 32
Petersson, J., 8

Q
Qatar Foundation, 126
Querin, O. M., 53

R
Rieser, J., 15
Robert McNeel, 119, 139
Rong, Y., 38, 44, 48–52
Rössl, C., 87
Rozvany, G. I. N., 64

S
Saha, P. K., 15
Sakai, N., 54
Sandberg, M., 15
Sano, K., 54
Sasaki, M., 2, 4, 19, 126
Sasaki, T., 54
Seidel, H.-P., 87
Seong, H. K., 82
Shao, Z., 54
Shen, W., 39, 124, 141
Shim, H. S., 59
Shi, T., 46
Shobeiri, V., 53–55
Sigmund, O., 1–4, 8, 15, 42, 46, 53, 63, 84
Sorkine, O., 87
Sounas, D., 54
Spangenberg, J., 15
Stam, J., 139
Steven, G. P., 2, 70, 87
Sun, Z., 54
Suresh, K., 63

T
Takeishi, A., 54
Takeuchi, T., 2
Takezawa, A., 82
Tian, K., 14
Toader, A. M., 69

Tortorelli, D. A., 46
Tran, P., 77
Tsuboi, Y., 2

U
Ueki, T., 2

V
Van den Berg, N., 129
Volk, M., 15
Voronin, A., 21

W
Wang, B., 14, 110
Wang, C., 31
Wang, G., 14
Wang, J., 39, 141
Wang, Q., 32
Wang, S., 54
Wang, X., 54
Weihong, Z., 60
Wei, P., 31, 135
White, D. A., 21
Wolfartsberger, J., 118

X
Xia, L., 46
Xia, Q., 46
Xie, Yi Min, 1, 7, 37, 59, 81, 107, 133
Xie, Y. M., 2, 4, 7–10, 12, 21, 22, 31, 34,
    38–40, 42–45, 48–55, 61, 62, 65, 66,
    70, 71, 77, 87, 88, 97, 98, 101, 108,
    109, 111, 114, 115, 118, 119, 121,
    123–125, 127–129, 133, 135, 139,
    141
Xiong, Y., 8, 10, 12, 31, 44, 46, 69, 77, 124
Xu, S., 110
Xu, T., 44, 77

Y
Yamada, T., 40
Yan, F., 135
Yang, J., 51
Yang, K., 7, 18–20, 39, 41
Yan, X., 8–10, 12, 31, 44, 77
Yan, Y., 51
Yao, S., 31, 69
Yao, Y., 115
Yoo, J., 82

Yuan, P. F., 133
Yuksel O., 15

Z
Zelickman, Y., 59, 60
Zhang, L.-Y., 42–44, 46, 51
Zhang, W., 31, 32, 42
Zhang, Z., 133
Zhao, L., 133
Zhao, Z.-L., 4, 7, 8, 18, 31, 38, 39, 41–44, 48–52,, 115, 124, 139

Zhou, J., 32
Zhou, M., 64
Zhou, Q., 7, 39, 141
Zhou, S., 7, 8, 39
Zhou, X., 133
Zhou, Y., 9, 14, 39, 42, 110, 141
Zhu, B., 40
Zhu, H., 141
Zhu, Y., 31, 32
Zimmermann, M., 15
Zuo, T., 31
Zuo, Z. H., 66

Index

A
Additive manufacturing, 15, 31, 44, 143
Aesthetics, 1, 2, 4, 16, 38, 107, 110, 115,
    120, 133–135, 137, 139, 142, 143
Architect, 7, 16, 19, 37, 42, 43, 115, 126

B
Bi-directional Evolutionary Structural
    Optimization (BESO)
  method, 10, 12, 14, 15, 17, 21–26,
    28–30, 34, 39, 42, 44, 64, 74–76, 98,
    113, 115, 120, 122, 123, 127,
    133–135, 141
  parameters, 115, 124, 126
Boundary condition, 44, 45, 61, 67, 85–87,
    91, 92, 97, 99, 102, 124
Bridge, 20, 21, 38, 39, 41, 50–52, 70–72,
    74, 75, 81, 110, 115, 133–140
Brush stroke
  hardness, 109
  opacity, 109
  radius, 109

C
Cavity, 30, 33, 34
Chair, 110, 115, 116, 133, 135, 139,
    141–144
Client, 1, 2, 4, 7, 8, 37, 38, 56, 134
Column, 16, 17, 27–30, 59, 60, 89–91, 137,
    138
Combination, 68, 86, 88, 90, 92, 99, 108,
    110
Compliance
  normalized, 10, 16

Computational workflow, 66, 67, 85, 98,
    99, 113–115, 119
Computer-aided design, 66, 119, 139
Construction, 16, 38, 119, 133, 134, 136,
    137, 139, 140
Convergence, 25, 28, 65, 66, 84, 95
Creativity, 1, 4, 55, 129

D
Design
  architectural, 1, 3, 4, 39, 107, 133
  child, 111–114, 116, 117
  conceptual, 4, 38, 135, 137, 138
  diverse and competitive, 2, 5, 7–9, 12,
    14, 17, 18, 24, 27, 34, 103, 118, 134,
    135
  domain, 1, 3, 5, 7, 9, 10, 12–14, 16–18,
    22, 27, 29, 30, 33, 37, 38, 41–55, 61,
    62, 67, 74–76, 78, 87, 121, 124, 126,
    135–137, 139, 143
  existing, 7, 18, 110
  exploration cycle, 118–122, 126–128
  freedom, 3, 4, 88, 119
  parent, 111–114, 117
  precedent, 19–21
  space, 3, 16, 17, 28, 37, 39–41, 48, 77,
    84, 115, 119, 136
  structural, 1, 3, 4, 42, 47, 55, 61, 62, 67,
    74–76, 78, 81, 82, 88, 104, 107, 115,
    124, 129
Dome, 9–14, 25–27
3D printing, 44, 143

E
Engineer, 2, 7, 37, 38, 47, 136

Evolutionary Structural Optimization (ESO), 2, 19, 71, 73, 76

F
Filter radius, 8–12, 14–16, 23, 27–30, 65, 120
Finite element analysis, 55, 61, 84, 142
Fused filament fabrication, 143

G
Genus, 110, 111, 117
Globally optimal, 1, 2, 7, 88, 90, 92, 104, 134

I
Integer, 66, 98, 113, 117
Interaction
 human–computer, 4, 5, 107, 108, 113, 120, 129, 143
Interior-point, 96, 104

L
Load
 condition, 1, 10, 12, 15, 18, 21–23, 26, 28, 40, 53, 77, 81–83, 85, 91, 98–100, 103, 104, 108, 119, 138, 139
 deterministic, 82
 direction, 24–26, 45, 82, 84, 85, 89, 93, 101–104
 location, 3, 5, 81–93, 97, 101–104
 magnitude, 5, 81, 82, 84, 87, 93, 94, 96–100, 104
 probabilistic, 82
 uniformly distributed, 14, 38, 39, 41, 46, 50, 86, 115, 119

M
Matrix
 flexibility, 89, 96
 stiffness, 61, 62, 64, 82, 89
Mechanical testing, 44, 45
Mixed reality, 129
Multi-material, 38, 133, 135
Multi-solution, 113–115

N
Non-design domain, 9, 12, 13, 16, 29, 30, 39–42, 50, 76, 87, 99, 108, 115, 119, 121, 126, 135, 137–139, 143

Non-linearity, 55
Non-reciprocity, 37, 54

O
Optimality criteria, 59, 63, 74, 84, 95, 104
Optimization
 algorithm, 4, 7–9, 22, 60, 74, 76, 88, 129
 constraint, 1, 10, 33, 34, 40, 45, 46, 61, 63, 66, 71, 94–96, 98, 101, 123, 137
 load, 95, 98–100
 method, 8, 14, 15, 23, 25, 34, 38, 48, 51, 52, 68, 70, 71, 74, 76–78, 88, 104, 107, 108, 118, 120, 129
 parameter, 7, 67, 87, 99, 120, 121
 procedure, 84
 process, 4, 7, 17, 21, 23–28, 30, 33, 39–43, 50, 52, 59, 60, 63, 66, 67, 74, 76, 82, 84, 95, 99, 101, 107, 108, 110, 111, 113, 115, 120, 123, 124, 143
 simultaneous, 5, 61, 62, 66–68, 70, 71, 74, 76, 78, 81, 85, 87, 88, 90, 93, 98–101
 stress, 46, 47
 support, 76, 77

P
Passive void domain, 41
Pattern
 drawn, 108–112, 115–117, 143
 geometric, 14, 42, 43
 preferred, 37, 39, 108, 110, 121–129, 143
Penalizing, 7, 18–21, 23, 24
Permutation, 90
Perturbation
 load, 24, 25, 27, 30, 31
 region, 28–30
 support, 27–29, 31
Practical application, 1, 2, 12, 16, 74, 90, 96, 126, 143
Preference
 aesthetic, 1, 2, 16, 107, 110, 134, 143
 artistic, 112
 subjective, 4, 5, 107, 108, 112, 117–121, 124, 128, 129, 139, 143

Q
Quadratic programming, 96, 104

R
Random, 21–25, 29, 30, 76, 103, 113, 117

Rendering, 12, 43, 44, 60, 127, 129, 134, 135, 142
Root
  fibrous, 51, 52
  taproot, 51, 52

S

Sensitivity
  analysis, 83, 84, 94
  number, 17, 64–66, 108, 110, 112, 113, 120
Shell, 9, 11–13, 25, 42–44, 67, 76, 77, 90, 92, 96, 99, 101, 103
Similarity
  geometric, 124, 126, 129
Smoothing, 119, 120
Software
  Abaqus, 61, 66
  Ameba, 141
  Grasshopper, 66
  Rhino, 66, 119, 121, 124, 126, 139, 142
Solid Isotropic Material with Penalization (SIMP), 69, 74–76, 98–100
Stiffness, 8, 17, 30, 45–47, 54, 59, 61, 62, 64, 71, 73, 78, 81, 82, 89, 98, 102, 110, 121, 122
Stress, 34, 45–47
Structural
  analysis, 84
  complexity, 7
  efficiency, 8, 21, 38, 76, 107, 108, 110
  element, 61, 62, 64, 66
  performance, 1–5, 7, 8, 10, 12, 13, 15, 16, 27, 31, 34, 39, 41, 42, 44, 51, 59, 60, 76, 78, 81, 86, 88, 104, 107, 108, 110–112, 115, 117, 120, 121, 124, 126–128, 133, 134, 139, 142, 143
  response, 3, 81, 90
  topology, 5, 37, 48, 59, 61, 64, 66–70, 74, 76–78, 81, 85, 87, 88, 90, 93, 98–101, 120, 124
Structure
  continuum, 34
  discrete, 34
Subjective
  preference, 4, 5, 107, 108, 112, 117–121, 124, 128, 129, 139, 143
  score, 107, 108, 112, 113, 116
Support
  condition, 1, 3, 7, 8, 18–21, 24, 38–42, 47, 49, 50, 59, 71, 78, 88
  element, 61–64, 66–72, 74, 76–78, 88
  fixed, 9, 119, 121, 126, 139
  material, 15, 31, 63, 78
  pin, 23, 76
  roller, 71, 77
Symmetry, 27, 70, 86, 103, 135–137

T

Topology optimization, 1–5, 7–10, 14–16, 21–23, 25, 30, 31, 34, 38–44, 46–49, 51–55, 59, 69, 74, 82, 87, 90, 94, 98–100, 104, 107, 108, 110–115, 118–122, 124, 127, 129, 133–136, 138, 139, 141–143
Tree, 85, 86, 88–90, 101–103
Tunnel, 30, 32–34

V

Virtual reality
  sculpting, 107, 118, 119, 121, 122, 124–127
Visualization, 108
Void, 21, 22, 30, 41, 43, 47, 49, 64, 66, 112, 115, 123
Volume fraction
  local, 10, 11

The manufacturer's authorised representative in the EU is Springer Nature Customer Service Centre GmbH, Europaplatz 3, 69115 Heidelberg, Germany. If you have any concerns regarding our products, please contact ProductSafety@springernature.com

Printed and bound by CPI Group (UK) Ltd, Croydon, CR0 4YY
26/03/2026
02078941-0010